Construction and Constraint

Studies in Science and the Humanities
from the
Reilly Center for Science, Technology, and Values

Volume I

CONSTRUCTION AND CONSTRAINT
The Shaping of Scientific Rationality

ERNAN McMULLIN, EDITOR

UNIVERSITY OF NOTRE DAME PRESS
NOTRE DAME, INDIANA

Library of Congress Cataloging-in-Publication Data

Construction and constraint.

(Studies in science and the humanities; 1)
Papers presented at a conference held at the University of Notre Dame in 1986.
    1. Science—Philosophy—Congresses.
2. Rationalism—Congresses. I. McMullin, Ernan, 1924-. II. Series.
Q174.C668   1988   501   87-40343
ISBN 0-268-00764-0
ISBN 0-268-00765-9 (pbk)

CONTENTS

FOREWORD

Modernity's legacy from the eighteenth century European Enlightenment has proven to be a deeply troubling one. The outstanding thinkers of this movement were unremitting in their condemnation of the many forms of prejudice, myth, and superstition which they held responsible for the unsatisfactory social and material conditions of their time. Yet these same individuals evinced a remarkable faith in human rationality and in its handmaids, the "new sciences," believing that once unfettered, human reason would not only conquer the physical world but usher in an unprecedented era of justice, equality, freedom and happiness.

Intimate acquaintance with the social and political history of the nineteenth and twentieth centuries is scarcely required to realize that the Enlightenment's optimistic prognosis for Western society has not been borne out. Assessments of this discrepancy, and reflection on its meaning, have preoccupied scholars as diverse as Nietzsche and Max Weber in the nineteenth century, and Heidegger and Jürgen Habermas in our own. In "The Rage Against Reason" in this volume, Richard Bernstein highlights one of the central themes in this critical tradition, the idea that the Enlightenment failed to envision what the real consequences of seeking to rationalize narrowly every dimension of human experience must be. But despite widespread skepticism about the benign role of rationality in politics and the shaping of society, one plank in the Enlightenment program has seemed secure, namely, its conviction that science embodies the capacity to transform our understanding and control of nature, and that it possesses this capacity by virtue of exemplifying human rationality in its purest form.

The idea that science as it has evolved in Western culture acquired a unique, and uniquely effective, mode of knowing and explaining is one that still exerts a powerful attraction. As recently as thirty years ago, philosophers of science were nearly unanimous in holding that the mature physical sciences embody powerful principles of method and reasoning which virtually guarantee that knowledge claims receiving the endorsement of a scientific community are free from any taint of personal bias, ideology, or gratuitous metaphysical prejudice. Thus the task such philosophers set for themselves was not that of *criticizing* scientific rationality, but rather that of seeking to *understand* its precise nature in a more transparent way.

Those aware of developments in the philosophy of science since the publication of Thomas Kuhn's *The Structure of Scientific Revolutions* just twenty five years ago will be struck by how dated this conception of science now appears to be. During the last quarter century every aspect of this "received view" of science has been subjected to serious criticism: that science is a unique mode of inquiry; that its methods embody a timelessly valid mode of rationality; that its goals and criteria of success are insulated from social and ideological pressures; that its accepted theories reflect only the structure of reality and not human biases grounded in struggles for power and privilege within specific historical and cultural settings.

Of course it is scarcely news that scientific theories, and the claims about nature they embody, change quite radically over time. And this sort of change is by no means incompatible with the view that enduring epistemic values of the scientific community might be invoked to explain the changing *content* of science. It is a different matter altogether to hold that those very second-level principles which shape and define the practice of science not only undergo transformation themselves, but do so in ways that cannot be satisfactorily explained by appeal to any deeper, nonrelativized conception of rationality. Yet precisely this conclusion has been the burden of much recent work in the history and sociology of science.

Is science merely a social construct? Are the goals and epistemic values animating science in a particular place and time to be understood primarily as a reflection of cultural and socio-political factors normally thought to be irrelevant to the practice of science? Even if socially rooted conventions are implicated in what passes for scientific rationality, does it follow that there are no normative epistemic constraints of a nonrelativist sort on the social shaping of science? Such questions are as complex and difficult as they are important. The need for reflective assessment of current debates about the concept of rationality and its role in science has become an increasingly urgent one.

To address this concern, a conference was held at the University of Notre Dame in April 1986 that took as its theme "The Shaping of Scientific Rationality." Papers presented at that conference by a notable and diverse group of philosophers comprise the substance of this volume. The essays gathered here display a broad range of perspectives on the increasingly polarized debates about the meaning and role of rationality in science. An edited transcript of the panel discussion that concluded the conference highlights critical areas of disagreement among the speakers, as well as issues on which consensus had begun to emerge.

The occasion for the conference was a special one: the inauguration of one of Notre Dame's most widely respected faculty members, Ernan McMullin, as O'Hara Professor of Philosophy. Those asked to speak at the conference were invited not only for their acknowledged expertise on the topic of rationality, but because each had, in his or her own way, a long and close association with the Department of Philosophy at Notre Dame, as well as with the recipient of the new chair.

The conference also was the first to be sponsored in part by the recently established Reilly Center for Science, Technology, and Values. The welcome gift of a substantial endowment for such a center from a generous university alumnus sets the stage for a period of renewed growth and activity in the History and Philosophy of Science at Notre Dame. High among the Center's priorities is sponsorship of further conferences on issues at the cutting edge of scholarship in the various fields of science studies. In addition, the Reilly Center is pleased to announce a new publication series through the University of Notre Dame Press: "Studies in Science and the Humanities from the Reilly Center" in which *Construction and Constraint: The Shaping of Scientific Rationality* appears as Volume One.

VAUGHN R. McKIM
Associate Director
Reilly Center for Science,
Technology, and Values

CONTRIBUTORS

RICHARD J. BERNSTEIN is professor of philosophy at Haverford College. He is the author of *Praxis and Action* (1971), *The Restructuring of Social and Political Theory* (1976), *Beyond Objectivism and Relativism* (1983), *Philosophical Profiles: Essays in a Pragmatic Mode* (1986), and editor of *Habermas and Modernity* (1985).

RICHARD FOLEY is professor of philosophy and chairperson of the Department of Philosophy at the University of Notre Dame. He specializes in epistemology; his book, *The Theory of Epistemic Rationality* has just appeared (1987). Some recent articles include "Dretske's 'Information-Theoretic' Account of Knowledge" (1987), "Evidence as a Tracking Relation" (1987), "Is It Possible to Have Contradictory Beliefs?" (1986), "What's Wrong with Reliabilism?" (1985), and "Epistemic Luck and the Purely Epistemic" (1984).

GARY GUTTING is professor of philosophy at the University of Notre Dame. He is coauthor of *The Synoptic Vision: Essays on the Philosophy of Wilfrid Sellars* (1977), co-editor of *Science and Reality* (1984), editor of *Paradigms and Revolutions: Applications and Appraisals of Thomas Kuhn's Philosophy of Science* (1980), and author of *Religious Belief and Religious Skepticism* (1982). His just-completed *Michel Foucault and the History of Reason* will appear in 1988.

xi

MARY HESSE is emeritus professor of history and philosophy of science at Cambridge University. She is the author
of *Forces and Fields* (1961), *Models and Analogies in Science*
(1966), *The Structure of Scientific Inference* (1974), *Revolutions
and Reconstructions in the Philosophy of Science* (1980), and most
recently coauthor, with Michael Arbib, of *The Construction of
Reality* (1986).

THOMAS MCCARTHY is professor of philosophy at North-
western University, having previously taught at Boston University and the University of Munich. He is the author of
The Critical Theory of Jurgen Habermas (1978) and co-editor
of *Understanding and Social Inquiry* (1977), and *After Philosophy* (1987). The series *Studies in Contemporary German Social
Thought*, which he edits for MIT Press, now includes some
three dozen volumes.

ERNAN MCMULLIN is O'Hara Professor of Philosophy and
Director of the Program in History and Philosophy of Science at the University of Notre Dame. He edited *The Concept of Matter* (1963); *Galileo Man of Science* (1967); *Evolution
and Creation* (1985), and is the author of *Newton on Matter
and Activity* (1978).

RICHARD RORTY is Kenan Professor of Humanities at the
University of Virginia. He is the author of *Philosophy and the Mirror of Nature* (1979), *Consequences of Pragmatism* (1982), and the forthcoming *Contingency, Irony and Solidarity* (1988). He also edited *The Linguistic Turn* (1967) and
co-edited *Philosophy in History* (1984).

THE SHAPING OF SCIENTIFIC RATIONALITY: CONSTRUCTION AND CONSTRAINT

Ernan McMullin

1. Introduction

The field we call "philosophy of science" has changed dramatically since I first started working in it more than thirty years ago. At that time the main preoccupations appeared to be *logical* ones. Philosophers saw natural science as a highly specific mode of knowing and of explaining; many said the *ideal* mode of knowing and of explaining. And so they asked questions like: how is hypothesis related to evidence in science? What is the logical form of scientific explanation? The presumption was that these questions could be answered in a formal and more or less definitive way. There was a logic underlying the methods of validation and of explanation in science, and the task of the philosopher was to disengage this logic once and for all.

The gradual collapse of this presumption in the 1950s and 1960s was surely one of the more significant developments in philosophy in our time. After all, this presumption was not just a modern aberration, an invention of the

Enlarged version of the Inaugural Address in the O'Hara Chair of Philosophy at the University of Notre Dame, April 11, 1986. I am indebted to Alasdair MacIntyre, Larry Laudan, and Dudley Shapere for their helpful comments on the first draft of this paper.

1

logical positivists; it had been foundational in most philosophic thinking about science, back through Mill and Kant to Descartes and Bacon and ultimately to Aristotle. Its abandonment marked a break between what might be called "classic" philosophy of science and its as yet unnamed successors.

2. Kuhn and the Stability Thesis

The most influential early exponent of the new way, Thomas Kuhn, shifted attention from matters of validation and explanation to the manner in which science *changes* over time. His now-familiar distinction between normal science and revolutionary science separates two sorts of change: one the cumulative solving of problems set by an accepted paradigm and the other the abrupt replacement of one paradigm by another. The language of gestalt shift and conversion that Kuhn used to describe the revolutionary mode of change was deliberately chosen to distance his account from those of his predecessors. What he wanted to underline was that the transition from one paradigm to another could not be brought about by force of logic alone. Though reasons played a part, even a crucial part, in motivating such transitions, they were not coercive; some would be persuaded by them, others would not be.

Most philosophers found this challenge to older orthodoxies rather shocking at first; some indeed still do. But my concern here is not with the merits of this controversy. My aim is to note a conservative aspect of Kuhn's analysis. He focuses mainly on changes in first-level science: in theories, in instrumentation, in textbooks, and so on. He does not say much about changes in scientific rationality itself, in the second-level principles according to which the scientific undertaking itself would be directed. Indeed, he appears to suppose that in what he calls the "mature" sciences there is a common rationality, marked by such shared values as predictive accuracy, consistency, simplicity, and so on. These "do much to provide a sense of community to natural scientists as a whole," he says in the Postscript to the second edition of *The Structure of Scientific Revolutions*.[1]

Even though they are shared, they may however differ in their *application;* different individuals will understand, or will weight them differently and so will make different judgments as to whether a particular theory is plausible or not. These differences occur even within the group of those who share a single paradigm:

> Though values are widely shared by scientists and though commitment to them is both deep and constitutive of science, the application of values is sometimes considerably affected by the features of individual personality and biography that differentiate the members of the group.[2]

But they are even more in evidence among those who defend different paradigms:

> What was for Einstein an insupportable inconsistency in the old quantum theory, one that rendered the pursuit of normal science impossible, was for Bohr and others a difficulty that could be expected to work itself out by normal means.[3]

Here the difference in regard to what qualifies as "rational" prevented agreement in regard to the acceptability of the new paradigm.

One might have expected that Kuhn would have made more of differences of this latter sort, because they support, much better than any other consideration he brings forward, his thesis about the intractability of paradigm disagreements. If two people do not even agree on what constitutes a good reason in a scientific argument, it is obvious that differences between them in regard to the acceptability of a proposed new paradigm are going to be difficult to bridge. Yet Kuhn prefers to stress first-level differences that lead scientists to view the world differently; where one sees a falling stone, the other sees a pendulum, and so on. What separates the protagonists here is not so much a difference in regard to the principles of scientific argument as a difference in the ways of articulating experience linguistically. And many of the examples he gives of paradigm-change (e.g., those brought about by the discoveries of the Leyden jar and of X-rays)[4] do not involve the principles of scientific rationality to any significant extent.

Kuhn returned to the issue of what he calls "value invariance" in an essay in *The Essential Tension*. He notes that in this essay:

> I have implicitly assumed that, whatever their initial source, the criteria or values deployed in theory choice are fixed once and for all, unaffected by their participation in transitions from one theory to another. Roughly speaking, but only very roughly, I take that to be the case. If the list of relevant values is kept short . . . and if their specification is left vague, then such values as accuracy, scope, and fruitfulness are permanent attributes of science. But little knowledge of history is required to suggest that both the application of these values and, more obviously, the relative weights attached to them have varied markedly with time and also with the field of application. Furthermore, many of these variations in value have been associated with particular changes in scientific theory.[5]

Kuhn says that there are specifiable criteria of rational theory-choice that are "permanent attributes of science." These are "fixed once and for all," very roughly speaking, at least; they are by and large "unaffected" by paradigm change. Once the level of paradigmatic science has been reached, further changes involve only the "application" of these values. Scientific rationality, as represented by the values involved in theory-choice, is thus fairly stable within revolutionary as well as normal stages in science. This is not, I dare say, the position one might expect someone to adopt who has so often been criticized for undermining the rationality of science.

And, it must be admitted, it is not consistently maintained. There are passages in *SSR* where he appears to assume that paradigm difference *necessarily* involves disagreements in regard to "evaluative procedures," "rules," "standards," disagreements sufficiently deep to deprive the protagonists of the common methodological ground that adjudication of the difference would require.[6] When he discusses the incommensurability of competing paradigms, the first reason he gives for the disagreement between the proponents is that "their standards or their definitions of science are not the same."[7]

In this view, a change of paradigm would involve a change of "standards." But what motivates this latter change is not an appreciation that the new set of standards is, of itself, better but rather that the paradigm of which it is a part is better. While stressing that the change is never a logically coercive one, Kuhn reminds his reader that there are "reasons" (connected mainly with the growing awareness of anomaly in the older paradigm) for adopting the new paradigm. The reasons do not bear on the changes in rationality directly; he seems to suggest, indeed, that these are secondary and that they ought not be regarded as progressive. Standards abandoned in Newtonian science, for example, may once again be resurrected in the science of Einstein.[8]

Apart from this suggestion that changes in rationality (i.e., in the norms of what counts as "good" science) are ancillary to changes in paradigm and are thus neither separately justifiable nor permanent, Kuhn has little to say about how and why second-level changes occur and what significance they have. His inclination, as we have seen, is more often to minimize such changes by suggesting that some very general values employed in theory-assessment remain the same. Of the two theses regarding rationality that we have identified with classical philosophy of science, the logicality thesis (assimilating the rationality of science to that of a logical system) and the stability thesis (taking rationality to be more or less invariant over time), Kuhn rejects the first but is less opposed to the second than one might have expected.

Since the positivists accepted the logicality thesis, it was natural that they should also believe that scientific rationality does not undergo substantial development. Scheffler put it very concisely:

> Underlying historical changes in theory, there is moreover a constancy of logic and method which unifies each scientific age with that which preceded it and with that which is yet to follow. Such constancy comprises not merely the canons of formal deduction, but also those criteria by which hypotheses are confronted with the test of experience and subjected to comparative evaluation. We do not, surely, have explicit and general formulations of such criteria at the present time. But they are embodied clearly enough in scientific practice to enable communication and agreement in a wide variety of specific cases. Such communication and consensus indicate that there is a codifiable methodology underlying the conduct of the scientific enterprise.[9]

Philosophers of science in the Popperian tradition have continued to defend this notion of a single codifiable methodology underlying the entire scientific enterprise. In a recent essay, Elie Zahar argues that "methodology relates to the sciences somewhat in the same way as logic relates to mathematics."[10] Mathematicians got their proofs right, long before strict logical formalizations of those proofs were available. "We have thus good reasons for supposing that, even in those early days, mathematicians had an intuitive understanding of the structure of first-order proofs." Likewise, scientists have been getting it right for a long while in a similarly intuitive way. Indeed, Zahar takes it to be a presupposition of Lakatos' methodology of scientific research programs that:

> the presystematic methodology instinctively used by scientists in judging individual achievements did not change very much over the centuries, say from the time of ancient Greece

until the present day. MSRP does not, of course, presuppose absolute stability, but it nonetheless implies that deviations from its norms have been in the nature of local fluctuations very different in magnitude from large scale scientific revolutions. . . . Many people have a strong feeling that intuitive methodology, like intuitive logic, has been largely stable.[11]

Zahar believes that his "stability thesis," as he calls it, may be rooted in our "biological make-up," which remains stable in the face of social and cultural change. In this view, the methods of science are warranted on intuitive grounds, just as the rules of deductive logic are. We have developed these powers of intuition in the course of evolution of human intelligence; their reliability is attested to by the role they would presumably have played in aiding human survival. This may be called the *logicist* account of the shaping of scientific rationality.

3. Feyerabend: The Challenge to Naturalism

The stability thesis has been under fire now for some time. Feyerabend's attack on it was a celebrated one:

It is clear, then, that the idea of a fixed method, or of a fixed theory of rationality, rests on too naive a view of man and his social surroundings. To those who look at the rich material provided by history, and who are not intent on impoverishing it in order to please their lower instincts, their craving for intellectual security in the form of clarity, precision, "objectivity," "truth," it will become clear that there is only *one* principle that can be defended under *all* circumstances and in *all* stages of human development. It is the principle: *anything goes.*[12]

The point of Feyerabend's "epistemological anarchism," as he terms it, is not to reject rationality entirely but to insist that no rule in science is absolute, that "there are always circumstances when it is advisable not only to ignore the rule, but to adopt its opposite."[13] In *Science in a Free Society* he gets quite upset with the critics of his earlier work *Against Method* who took him to be rejecting rules and standards entirely. An anarchism as extreme as this would, he says, be naive and unworkable:

> I argue that all rules have their limits and that there is no comprehensive "rationality." I do not argue that we should proceed without rules and standards.... I suggest a new *relation* between rules and practices. It is this relation and not any particular rule-content that characterizes the position I wish to defend.[14]

This may, he says, sound like a "naturalism" which would derive its rules and standards solely from an analysis of the scientific tradition itself. But this approach, which he associates with Lakatos, will not work (he claims) because there has been a multiplicity of traditions and there is no way to distinguish the properly "scientific" one without begging the question. The naturalist is forced to assume that it has:

> been already established that modern science is superior to magic, or to Aristotelian science, and that it has no illusory results. However, there is not a shred of an argument of this kind. "Rational reconstructions" take "basic scientific wisdom" *for granted,* they do not *show* that it is better than the "basic wisdom" of witches and warlocks. Nobody has shown that science (of "the last two centuries") has results that conform to its own "wisdom" while other fields have no such results.[15]

This is a challenge to which I will return. Meanwhile, Feyerabend has a further difficulty for the naturalist. How is he to deal with "cataclysmic developments," what Kuhn called "revolutionary changes," in science itself? "New paradigms," it would seem, bring in a "new rationality."[16] And if this is so, to what standard can the naturalist turn? To justify, for example, the switch from Aristotelian science to Newtonian science in naturalistic terms:

one must either show that at the time in question the Aristotelian methods did not reach the Aristotelian aims or that they had great trouble reaching them while the "moderns," using modern methods, experienced no such difficulties relative to *their* aims, *or* one must show that the modern aims are preferable to the Aristotelian aims.[17]

This is a perceptive diagnosis. Feyerabend is convinced, however, that naturalists cannot make good on any one of these possible strategies. Since they cannot identify in properly naturalist categories the motives that led people to prefer the new science over the old one, they have no reason ultimately for asserting that the "professional ideology of modern science" which now "reigns supreme" is better than that of the Aristotelians. To explain the switch to Newtonian science, he goes on, one must examine instead "the function of propaganda, prejudice, concealments and of other 'irrational' moves" ("irrational," that is, to the naturalist) in the gradual solution of the problems that at first prevented the acceptance of the new science of Galileo and Newton.[18] One can, in principle, explain the change in *these* terms, he believes, but of course, such an explanation would in no way justify the naturalist construal of the change in terms of a unique form of rationality termed "scientific."

Nor would it permit one to set Newtonian science *above* Aristotelian science, no more than in a broader context one could set science itself *above* magic; the fact that one of the two prevailed in each case does not give it a claim to *rational* superiority since the means by which victory was achieved would, in part at least, not qualify as "rational" in the only terms a naturalist could allow. Feyerabend can conclude, then, that although what we call science today does have what might be described as a form of "rationality" proper to it, this rationality is much weaker than is usually supposed and, furthermore, it is only one among a number of alternative forms of "rationality," no one of which can claim superiority. And in the end, standards are no more the arbiters of truth in science

than they are of beauty in art. A scientist or an artist does
not need:

> papa methodology or mama rationality to give him security
> and direction, he can take care of himself, for he is the
> inventor not only of laws, theories, pictures, plays, forms of
> music, ways of dealing with his fellow man, institutions, but
> also of entire world views, he is the inventor of forms of
> life.[19]

Construction without constraint, it would seem.

Yet there appear to be two rather different strands in
Feyerabend's thinking. The anarchist, on the other hand,
stresses *invention* and alleges that the attribute of rationality gets attached to a piece of creative science only after
the fact. There is a plurality of purported rationalities; if
advantage is claimed for one of these over the others, it
can only be because the strategems and tricks of the prop-
agandist are being employed. On the other hand, in *Science
in a Free Society* there are the elements of what might be
called a dialectical view. The naturalist is right, he says, to
insist that practice can alter reason. But the idealist (logicist, in our terms) is also right to claim that reason can
alter practice. Thus, reason and practice must be held to
interact, as long as this metaphor of interaction does not
mislead one into thinking that the two can exist separately.
(One is reminded of Kuhn's assertion that standards of
rationality change in consequence of replacement of the
paradigms of which they are a part.)

Feyerabend's conclusion is that the rationality of science is something like "a guide who is part of the activity
and is changed by it."[20] The main source of such practice-
induced change, he notes, is cosmology. "The standards
we use and the rules we recommend make sense only in a
world that has a certain structure."[21] Standards may have
to be modified or abandoned if the world proves not to
have the structure they presuppose. Of course, one might
take leave to doubt whether the kind of science he de-

scribes would be capable of discovering underlying structure to begin with. But more of that later.

What of the naturalist option that Feyerabend draws
upon and yet ultimately rejects? It is odd that he should
have taken Lakatos to represent this view, Lakatos of all
people. Feyerabend himself reminds us of the way Lakatos
reconstructs history to his own purposes. The MSRP is not
really derived from a descriptive account of scientific
practice; Lakatos makes use of this practice by way of
illustration of his methodology but he is constantly ready
to amend it, and to invoke the influence of "external"
factors, when it does not conform to the pattern he
expects of it.[22] There is much more of logicism, of "critical rationalism" to use Popper's phrase, in this than
there is of naturalism. There are others, however, who
would take on themselves much more readily the mantle
of naturalism.

4. The Stability Thesis and the "Strong Program"

Looking at the contemporary scene in philosophy of
science, the most evident challenge to the stability thesis
comes from the naturalist approach adopted by sociologists of science, particularly by those who subscribe to the
"strong program" in the sociology of knowledge. Since
science is a social construction, its standards, as well as its
theories, must be understood as social conventions reflect-
ing the culture which gives them authority. What makes
the standard a *standard* is its acceptance by a community,
and that acceptance can be understood in socio-political
terms like any other social act. The notion that we can
"see" the standard to be the "right" one, that we can, for
example, tell without reference to social consensus that
repeatability is a proper demand in an observational science,[23] is a mistaken one, based on an outmoded rationalist ("teleological") view of mind. There is no logic of
science if by 'logic' one means a set of norms whose warrant is something more than the contingent consensus of

the community of scientists at a particular time. Scientific theories are metaphors and like all metaphors are culture bound; the criteria by means of which they are assessed vary from field to field, as well as over science as a whole through time. They are validated within the community by techniques of socialization of a special sort (Kuhn is usually quoted here). Science enjoys no advantage on the grounds of an autonomous or partly autonomous rationality: "Hence its process of cultural transmission will be in no important respect different to those employed by other knowledge sources."[24] It sounds like a flat rejection of the stability thesis.

As in the case of Feyerabend, however, there is another strand to consider, a secondary one in most theoretical writings in the sociology of science to be sure, but one whose implications may be more far-reaching that at first sight they appear to be. Barry Barnes writes that even though scientific knowledge has to be understood naturalistically in terms of its cultural antecedents, it also can be said to be constituted by an interest in prediction and control. He is aware of the Kantian overtones of this theme in the writings of Habermas and discounts the transcendental status that Habermas attributes to it.[25] But he seems prepared to allow that predictive accuracy and the technical control that flows from it is in some sense an *invariant* of what has come to be called "science." Though the theories and even the procedures of contemporary science are the contingent products of twentieth-century European culture and for that reason may be replaced by something unpredictably different (just as Aristotelian science, the distinctive product of Greek culture, was replaced by the science of Newton), the demand for predictive accuracy will remain, and will presumably constitute a constraint on the larger forms that scientific rationality may take in future communities of scientists.

This last inference is not brought out by Barnes himself, nor by others in the Edinburgh tradition, but it seems fair to say that the goal of accurate prediction that they assign to science introduces something other than the mere *fact* of social consensus as warrant for claims about the most effective means to this goal. Agreement about the importance of repeatability in experimental work, for example, is explained not merely by referring to a consensus in the scientific community as cause but by noting the reason for the consensus, namely, the link between repeatability and predictive power.

David Bloor writes:

> Think of the feats of endurance that North American Indian males were said to undergo in order to be fully initiated warriors of their tribe. That theories and scientific ideas be properly adapted to the conventional requirements that are expected of them means, among other things, that they make successful predictions. This is a harsh discipline to impose on our mental constitution; but it is no less a convention.[26]

I would be inclined to reverse the last sentence, and say "Imposing this on our mental constitution may involve convention, but it is no less a harsh discipline."

5. Laudan's Reticulated Model of Change

One of the most severe critics of sociology of science, Larry Laudan, takes naturalism in a different direction. In his recent book *Science and Values* he argues for what he calls a "reticulated" model of the relationship between aims, methods, and theories in science. No one set of these elements ought be regarded as privileged or more fundamental than the others. Instead of thinking of them as three *levels,* with justification proceeding downwards from aims to methods to theories—Laudan calls this the "hierarchical model"—one must treat them as mutually dependent. Our factual beliefs or our theories can, for example, alter our views as to what methods are viable and, in consequence, what goals are attainable. The goals of science are not set in advance; in fact, different, even mutually incompatible, goals may satisfy the relatively weak constraints imposed by the reticulated model. "There is no single 'right' goal for inquiry," Laudan concludes, "because it is evidently legitimate to engage in inquiry for a wide variety of reasons and with a wide variety of purposes."[27] Yet not *any* goal will qualify; "there are plenty of purposive activities which nonetheless fail to meet our intuitive standards of rationality."

This last phrase has a "hierarchical" ring to it, as Laudan realizes:

> Before a purposive action can qualify as rational, its central aims must be scrutinized . . . to see whether they satisfy the relevant constraints. But beyond demanding that our cognitive goals must reflect our best beliefs about what is and what is not possible, that our methods must stand in an appropriate relation to our goals, and that our implicit and explicit values must be synchronized, there is little more than the theory of rationality can demand.[28]

Yet this "theory of rationality" is by no means vacuous. Because it requires, for example, that goals that prove unrealizable be abandoned, it has ensured that changes in scientific methodology of a progressive sort should occur. Laudan suggests, indeed, that many of the theories of science now current would fail to meet its "modest demands." His approach is thus not entirely "reticulational," single-level. Above the interlocking goals, values, theories, there appears to be a prior theory of rationality sufficient to enable one to certify some changes of standard as "rational" in the light of historical practice.

Still, he calls his argument a "Heraclitean" one: "Theories change, methods change, and central cognitive values shift."[29] How can one speak of overall *progress* in science then? The answer is simple, he says. We need not judge progress by referring to the aims held by the scientists themselves. Instead, determinations of progress are to be made "relative to our own views about the aims and goals of science." To ask whether science has progressed is to ask "whether the diachronic development of science has furthered cognitive ends that we deem to be worthy or desirable."[30] Relativizing claims of progress to our own "cognitive aspirations" still allows us a "robust notion" of progress. We must guard against any more "absolutist" a sense of progress than this, he warns, and concludes: "There is simply no escape from the fact that determinations of progress must be relativized to a certain set of ends, and that there is no uniquely appropriate set of those ends."[31]

This notion of progress does not sound nearly as "robust" as the one he defended in his earlier book, *Progress and Its Problems.* There he claimed to "exploit the insights of our own time about the *general* nature of rationality," thus "transcending the particularities of the past":

> For all times and for all cultures, provided those cultures have a tradition of critical discussion (without which no culture can lay claim to rationality), rationality consists in accepting those research traditions which are most effective problem solvers. . . . There are certain very general characteristics of a theory of rationality which are *trans-temporal* and *trans-cultural,* which are as applicable to pre-Socratic thought, or the development of ideas in the Middle Ages, as they are to the most recent history of science.[32]

On the face of it, a profound shift seems to have occurred. In the earlier work, Laudan argued that "the single most general cognitive aim of science is problem-solving"; to be "rational" is to "maximize the progress of scientific research traditions" in terms of problem-solving capacity.[33] In the later book, he claims that there is no single "uniquely appropriate" set of aims for science, and hence the best we can do is to relativize assessments of progress to the aims *we* believe in. But this is surely at odds with the problem-solving approach to the assessment of past science, which requires one to identify with the standards that constituted the problems *as* problems and the solution *as* solutions. Adopting the standards of a later time, as Laudan's notion of "reticulation" leads him to advocate, would in this context defeat the effort to reconstruct historical problem-solving sequences.

What may prompt these two very different approaches are different conceptions of how to construe science as "progressive." According to one, an episode from the history of science is said to be "progressive" if it exhibits progress in problem-solving against the norms of that day and of the discipline. According to the other, it is "progressive" if it qualifies as progressive under *our* standards, as later chroniclers. Laudan made a point of claiming in the earlier work that the specifics of scientific rationality are "partly a function of time and place and context,"[34] so that, for example, it could have been perfectly rational, at a time when religious doctrines were held to be relevant to the understanding of nature, for a scientist to take such doctrines into account in formulating his theories. But the rationality of such a move would have depended, in Laudan's view, upon the extent to which the theological doctrines themselves could have been construed as "rational," that is, as progressive in problem-solving terms.[35] Rationality was thus tightly tied to problem-solving; one could assess alternative traditions like magic, alchemy, astrology, by means of this criterion and eventually decide in favor of science in the conventional modern sense.

Let us assume that Laudan still implicitly subscribes to the older view that one can define a very general transtemporal aim of science in terms of problem-solving ability.[36] This would modify the "reticulational" approach in an important respect by postulating a general vantage point above the specific and changing aims of science sufficient to enable transtemporal assessments of progress to be made, assessments that are not relativized just to *our* view of what the aims of science should be. Such a vantage point would enable him to respond to Feyerabend's challenge to a naturalism that remains at the level of the descriptive: how, then, is one to set science off from an enterprise like magic? Having available to him a general "theory of rationality," a general aim of science, is what allows him to be normative. But at this point his view surely ceases to be "Heraclitean"; the reticulated model has been supplemented. Though the constraint may be a light one, it is nonetheless constraint.

6. Shapere: The Piecemeal Approach to Change

Dudley Shapere has drawn attention to the challenge posed by an uncompromising historicism in philosophy of science. In such a view:

> scientific change is not merely a successive alteration of sub-stantive beliefs occasioned by new discoveries about the world; scientific change and innovation extend also to the methods, rules of reasoning, and concepts employed in and in talking about science. Even the criteria of what it is to be a scientific "theory" or "explanation" change; and thus the notion that *theory* or *explanation* are "metascientific concepts," with meanings independent of scientific beliefs, is rejected.[37]

But if this is the case, he asks, how can standards of rationality themselves, without circularity, be held to undergo "rational" change? He rejects what he takes to be the usual responses to this question: the "presuppositionist" response, according to which there are higher-level criteria of rationality, themselves immune to alteration, in terms of which changes of lower-level rationality can be judged to be rational, and the "relativist" response, according to which there are no real grounds for supposing that changes in the criteria of science are themselves rational and hence no reason for supposing that there can be real progress in science.[38] Instead, he proposes that there can be "a chain of developments connecting the two different sets of criteria, a chain through which a 'rational' evolution can be traced between the two."[39] A "traceable relationship of changes introduced for reasons"[40] will suffice to show that the change is rational, he believes, even if the notion of "reason" itself changes along the way. Changes in the aims of science or in the criteria of rationality are linked to changes in our substantive beliefs about the world; aims and criteria have to be proposed and modified just as the lower-level theories themselves have to be. We not only learn, we learn *how* to learn.

What Shapere in particular opposes is what he calls "essentialism": the supposition that there are defining characteristics of science, themselves not open to revision in the light of discoveries about the world. Even the laws of logic may have to be modified in the light of scientific practice. Nothing is unrevisable, but change is gradual, "piecemeal." Progress in science "consists partly in sharpening the reasons for doubt," and the doubt can extend all the way up and down the line from the observational level to the most general principles of rational method. Of course, when doubt arises about some element on the basis of what we have "learned to take as well-founded reasons," we proceed in a definite order, questioning first the beliefs that are already suspect or are the least costly to give up and only gradually extending questioning upwards to the most general level. There are overtones of Popper and Lakatos in all this, but Shapere is allowing potential falsification to extend much further than they would have done, and it leads him to suggest that his views constitute "perhaps the first truly uncompromising empiricist philosophy ever proposed."[41]

But a question immediately suggests itself. If change extends to *all* levels, are there then *no* constraints on the kind of activity science might become? Could it, in principle, become something entirely unlike what it is today by way of piecemeal change? Might its goals, for example, at some remote time in the future, come to resemble the goals of the activity we now call football? Shapere requires that there be "reasons" for change at every step of the way, in order that the activity itself remain a rational one. Does this implicitly set limits to the scope of possible change? Might not a continuity of this sort force the acknowledgment of some very general kinds of beliefs or rules that are ultimately requisite to any activity that is to count as science?

When these questions were pressed[42] at a panel discussion of the paper we have been quoting, Shapere did not concede. The reasons for change are determined by the *content* of science at a given time, by its rules, methods, substantive beliefs and their interplay so that what counts as a "legitimate successor" will itself change, and these changes cannot be anticipated in advance:

> I give no general unchanging "notion" of legitimate succession, nor do I need to do so to assure continuity of the enterprise. . . . The continuity of science can be assured without assuming common essential features as long as there are chain-of-reasoning connections, and even what counts as a "reason" may change, so that the character of the chain may too.[43]

We have, he continues, no reason to think that science ever *will* become anything resembling football. But this is not sufficient grounds for the conviction that it *cannot*. Our present characterization of science cannot be converted into anything like a constraint on the forms that science may *ultimately* take. There is a "Platonic fallacy" underlying the assumption that there must be something in common across all instances of a term's use. Even if the constraint were to be specified in the most general way, how could it be shown to be *necessary* for science without invoking some kind of transcendental argument, an appeal to essence? We can no more anticipate the future forms that scientific rationality may take than Democritus could have anticipated modern atomic theory.[44]

In a later essay, Shapere returns to this issue. He defines two very general sorts of criteria for science; "success" describes how well a particular scientific account
does account for the items of its domain, while "adequacy" bears on such features as consistency, complete-
ness, and compatibility. These criteria themselves are the
product of a historical process and have undergone alteration along the way; once regarded as transcendent, they
have been "internalized" into the scientific process, "becoming subject to the very procedures of revision or rejection which they themselves helped define."[45] In this way,
the distinction between "levels" within science, or between science and "metascience," has been eliminated, or
at least blurred, since all aspects of science are now seen
to be open to test. This also implies that "no transcendental argument can establish the unalterably *a priori* status of
any claim," even about the most general features of science. The standards of success and adequacy have been
progressively refined in a "bootstrap" way that avoids circularity because of the mutual reinforcement between beliefs about the universe and the criteria of those beliefs.[46]

It is a matter of historically contingent fact that these
criteria have come to be what they now are, with all the
successes we now attribute to them. These successes have
"led to the establishment of an *ideal,* a normative principle for science . . . established precisely because it has
been fulfilled to a very high degree, even if not completely."[47] That ideal specifies that a scientific belief must
satisfy the by-now stringent conditions of success and adequacy that have evolved piecemeal through the long history of science. We had to learn how to learn.

But at this stage, what does it *mean* to say that we have
learned? This is the real issue. The notion of learning
suggests that there has been a positive achievement, that
we now *know,* at least approximately, not in any *a priori*
way but on the grounds of experience, what goals are
achievable in the sort of inquiry designated as "scientific," and by what means. At this point, however, Shapere

once again draws back. After saying that the criteria of success and adequacy have become "established" as an ideal for science, his fallibilism still leads him to qualify this in a crucial way. Though these criteria can be regarded as "constraints . . . which, because they have proved so fulfillable, have been adopted as goals or ideals, to be satisfied as fully as possible in any scientific situation," he warns that they remain "subject to further alteration or rejection in the light of new findings."[48]

Let us be clear what rejection of these ideals would entail. It would mean, for example, that success in accounting for the data would no longer be considered as a normative principle for science, that it would no longer be thought important for a theory bearing on the future to predict accurately. We can expect that the precise manner in which a theory "accounts for" the data will be further "refined" (to use Shapere's term) as time goes on. Refinement, however, is one thing; rejection is another entirely. It is not enough to say that we have no positive reason to suppose that rejection *will* happen, that these are "mere possibilities—only philosophers' considerations that are really no considerations at all."[49] To allow that it is possible in principle for predictive success to be set aside entirely in the "science" of the future has far-reaching repercussions here and now in terms of the attitude one should adopt to the "science" of the present.

Our conviction is that it is at least partially "correct," that its findings have some permanent worth, that its goals have been at least partially fulfilled. If it is in principle even *possible* that through a series of rationally justifiable changes, these goals might have to be entirely set aside, our convictions about the epistemic and ontological status of the best-supported theories in contemporary natural science would have to be substantially qualified. Shapere is uneasy about *any* definitive constraint on the future of scientific rationality because of the unwelcome suggestion of an essence known *a priori* that the idea of constraint conveys to him. Yet if we have learned that certain goals are fulfillable through the use of certain means, it sounds as though there are already quite definite constraints on long-range future possibility. If future "piecemeal" changes in rationality will themselves be rational, as Shapere is willing to prophesy, certain sorts of outcome can (it would seem) be excluded with confidence.

So much, then, for our review of the spectrum of positions that philosophers of science have taken up in regard to the historical character of scientific rationality. They range from Zahar at one end to Feyerabend (in some of his pronouncements, at least) at the other. Nearly everyone is willing to allow *some* kind of constraint on the future shapes scientific rationality may take. But there is very little agreement as to just what form that constraint should take and what its sources are. In the second half of this paper, I want to review this issue, taking my lead from the questions posed by the philosophic views I have been discussing. I shall look to the history of science for clues without assuming that science is something like a natural kind. Logicists and historicists can still agree that one must not attempt to prescribe a rationality for an activity as complex as natural science without first examining its actual practice. The questions I am particularly concerned to answer are two. First, does the rationality of science change over time, and if it does, for what reasons? Here I find myself eventually leaning to the left. Second, are there specific constraints on such change? Does science have any broadly defining characteristics? Here I tend to lean a little to the right. That should, on balance, leave me fairly close to the middle!

7. Some General Notions

We have lots of words to describe the most general features of human activity, but most of these are quite vague in everyday usage. It is important to clarify a few of them before we begin to look at the historical record of science. First, *goals* or *aims*. I take the goals or aims of a human activity to be the outcomes that prompt the agent to perform the activity. They are what the activity is intended to achieve, though of course they may fail to be realized in given cases. The goals of science are whatever the scientific community intends to achieve through scientific inquiry. Individual scientists obviously have all sorts of goals as they pursue their researches, like getting tenure or becoming famous. But when I speak in a general way about "the" goals of science, I mean those whose achievement would constitute success in the eyes of the community of scientists. Second, *methods*. In order to achieve these goals, certain methods or procedures must be followed. Methods are a means to an end. Their justification is that they serve what are taken to be the goals of science. They are instrumental, not ends in themselves. Methods are sometimes stated in the form of explicit *principles,* specifying what should be done or avoided, but more often they are learned as a skill of a tacit sort. Finally, *values.* Value-judgment is an important part of the methodology of science. Scientists assess measurements for their accuracy, experiments for their role in theory-testing, mathematical formalisms for their elegance, theories for their consistency or the like. Values in this context are characteristics which the community regards as desirable in the entity that is being assessed.[50]

When one speaks of the *rationality* of science, one is usually referring in a global way to the methods employed by scientists as well as the values they try to maximize in the course of applying these methods. It is in this sense instrumental; it is the means to whatever ends natural science as an activity aims at. It is not autonomous, not an end in itself.

This point needs emphasis among philosophers and mathematicians. When a formalism is proposed as a means of understanding some complex subject, the formalism can easily come to have a life of its own; it can come to be studied as an end in itself. Think of the rational mechanics of the eighteenth century, for example, or the logic of Russell and Whitehead's *Principia Mathematica* in our own day. Shapere reminds us of the damaging effects of *PM* when it was taken to *define* reasoning, instead of being regarded as a partial means of understanding a complex activity we were already performing. The seductive thing about formalisms is that they can quite legitimately be studied as ends in themselves, and this can be a very satisfying exercise. But if the original function of the formalism was to help us understand something other than the formalism itself, the properties of moving bodies, say, or the nature of inference, this function may all too easily get lost from sight. The danger is more real in logic than in physics, where the tie to the real order is more immediate. We all remember the so-called "paradoxes" of material implication which are paradoxes only if one assumes that the *PM* formalism is adequate to the task of conveying what is meant by 'if—then' in ordinary language.

The relevance of this to the topic of scientific rationality is perhaps obvious. Scientists make use of a variety of kinds of inference, and these can be at least partially formalized. These formalisms can be (and indeed have been) developed as ends in themselves. But the crucial question is: do they illuminate what the scientist is actually *doing?* And the danger is that to the extent they do not, this will be taken by the philosopher to be a deficiency on the part of the scientist instead of an inadequacy of his or her own formalism.

How and why do goals change? For two reasons especially. One is because, in their present form, they cease to seem worthwhile; the other is because they come to seem unattainable. These two factors, the relative *worth* of the goal to the person or to the community, and its *attainabil-*

ity, are what we need to take into account in tracing how goals change. Likewise, realization of a need or desire on the part of the community may lead to the formulation of a new goal, once it comes to seem attainable. Notice that goals are not beliefs, though they involve beliefs. They are not presuppositions. It may be in part because Shapere classes the goals of science as beliefs that he insists on the in-principle possibility that they could at some later time be entirely rejected in their present form. After all, he urges, if our most fundamental beliefs about the natural order have changed and will continue to change in science, how could goals of even the most generic kind remain the same?

It is obvious that changes in beliefs about the world may indeed affect the goals of inquiry. They are not irrevisable; they may be modified, sharpened, broadened. But the issues raised by the total revisability thesis go beyond such change. Can the goals of science be modified in small stages, in such a way that the initial goals and the final ones have nothing in common? Would one want to call this the "same" activity throughout? It sounds like the medieval puzzle about the raft that is rebuilt, plank by plank, until nothing of the original timber is left. I am not at all sure how activities are ultimately individuated, though I am inclined to think that a total change of goal over time should debar one from speaking of the "same" activity. But that is not where I want to rest my case; the matter of the determinacy of reference of terms like 'science', subtle and fascinating though it may be, is not why I think this is an important issue.

A word on how and why rationality might change. If scientific rationality is a means to the goals of science, then it will of course change if the goals change. But, more important, it will change if scientists discover more effective ways of realizing the goals. The question: how can changes in scientific rationality themselves be rational? is easy to answer, once it be noted that the reference of the term 'rational' shifts within the question. If one aims at certain goals, it is rational, reasonable, to attain these goals in the most effective way. The "reasonableness" we call on here is not the rationality of science but the more basic rationality that informs goal-directed human action. It is reasonable to modify the rationality of science if we can in consequence do better science, that is, attain the goals of science better by doing so.

8. A Little History: The Early Period

How might these disagreements about the shaping of scientific rationality be tested against history? A safe place to begin would be with Plato and Aristotle, on the grounds that there is an easily traceable continuity from them to present-day natural scientists, with each generation citing the previous ones, quarreling with them, improving on them. What were the goals of Aristotle's natural philosophy? They were principally two: first to provide an explanation of the world of nature, and second to make that explanation secure as eternal and necessary knowledge. The activity of explaining the general features of nature did not begin with him or even with the pre-Socratics. It goes back to the cosmogonies that are likely almost as ancient as humankind itself. What marks off the Greek moment of the story is the union between two goals not previously related: explanation and knowledge, construed as justified true belief. Lloyd speculates that the early Greek law courts and parliaments may have had something to do with this new demand for knowledge that could sustain challenge.[51] We do not know. What we do know is that the two streams flowed together in Aristotle to produce the first self-conscious account of what science, *epistēmē*, should be.

Notice how directly the rationality of Aristotle's science corresponded with the goals it was intended to serve. Explanation was to be achieved by a grasp of essence. Natures were to be interrelated by a schema of genera and differences. Change was to be understood in terms of the four causes, all of which were themselves rooted in essence. As knowledge, it took the simplest form possible: all inference was to be deductive, that is, entirely rule-governed, and the starting points were to be anchored by intuition (*epagōgē*), schooled by experience. Aristotle devoted a great deal of effort to the formalism; later generations would be fascinated by the elegance and intricacy of his theory of the syllogism. But he had little to say about the intuitions of essence on which the entire structure depended.

It was all too easy to reduce the rationality of his science to a logic which would serve at once as proof and as explanation. If our minds had had the powers of intuitive penetration that he assumed, and if the natural world were in fact made up of interlocking essences of the kind he postulated, then the rationality of science *would* indeed reduce, virtually, to logicality. What defeated this attractive but much too optimistic first attempt was that neither our minds nor our world proved to be of the sort that Aristotle had hoped they were. Note that this had to be discovered through the failures in *practice* of the Aristotelian proposal. There could well be a world in which this proposal *would* have worked; only the extensive experience of trying and failing could decide whether that was our world or not. There was nothing in the logic of syllogism itself, considered as a formalism, which would have helped decide whether or not such syllogisms could actually provide a science of nature.

The kind of understanding this science gave was not seen as being closely related to *prediction,* either as outcome or as test. To understand a nature was to know, among other things, how the entity possessing that nature would behave under normal circumstances. So there was an element of prediction there, if you like. But there was no attempt to *test* the understanding by means of systematic application; there was no need of test after all, since the knowledge of essence proceeded from an intuitive and self-certifying grasp of first principles. Even where Aristotle hinted that the quality of the scientific knowledge given (of the steady light of planets by contrast with that of stars, for example) was dependent on the quality of the explanation provided by the postulated middle term, nearness, there was no suggestion that prediction of new effects could be a result.[52]

Yet there was, of course, an activity of systematic prediction going on at that time, and for long before. One realm of invariable order that had been evident even to primitive people was that of the heavenly bodies and the many earthly effects dependent on their motions and influences. For more than a thousand years before Aristotle's day, the Babylonians had seen an omen-relationship between certain sorts of celestial appearances, principally lunar eclipses and the times of first and last appearance of various stars across the horizon, and the details of daily life in the empire. They had gradually developed a systematic record of lunar and planetary observations on the basis of which predictions could be made, using complex arithmetical functions. They did not (so far as we can tell) see this as in any way related to an *understanding* of the heavenly bodies or of the causes of their motions; were these not, after all, serving as the message-bearers of the gods? And the goal of their mathematical astronomy, which reached its zenith a century or two after Aristotle's death, was not, originally at least, prediction for its own sake but prediction in the service of a further goal, a practical knowledge of terrestrial outcomes.[53]

Observational astronomy was not nearly so advanced among the Greeks, but they did ultimately develop a mathematical astronomy, based on geometrical rather than arithmetical principles. Aristotle borrowed the scheme of Eudoxus to incorporate in his own science of nature, but it is not clear that this scheme could ever have been used for practical predictive purposes. Four centuries later, the great Alexandrian scientist, Ptolemy, made a determined attempt to construct an astronomy that would be at once predictive and explanatory but had to confess failure, and was forced to be content with a model of superb predictive power but virtually no explanatory virtue at all. Never was it so obvious as in Ptolemaic astronomy why prediction and explanation cannot be assimilated to one another.

There were other activities going on in the Greek world of the day that should not be forgotten. There was medicine; there were the crafts of agriculture, architecture, metallurgy, dyeing, and the like. All of these required some degree (sometimes a considerable degree) of practical knowledge, a knowledge that was handed on within families or within guild-groups. This knowledge was not derived from first principles but from centuries of experience and the day-to-day testing in terms of visible success and failure that goes with the practices of healing or of making objects responding to human needs. Though attempts were made to link medicine with the science of human nature, or to make use of optical principles to construct such devices as burning mirrors, these were few and untypical. There were social barriers to the merging of crafts with science, but more important, the distance between them as forms of knowledge was far too great to make any systematic efforts to unite them seem worthwhile.

The long period of relative neglect of natural science that followed reminds us that the goal of explanation or understanding of natural phenomena, taken in isolation, is unlikely to be pursued except under favorable social conditions. It requires leisure, it requires stability, it requires a supportive educational system, and much more. The goal of understanding simply for understanding's sake can to all intents and purposes be abandoned in times when human survival is threatened. Important also in explaining the long eclipse of scientific inquiry was the fact that an alternative and much more reliable form of knowledge seemed to be available in the Bible. The goal of secure knowledge which had led Aristotle to his notion of *epistēmē* was more likely among Christians and, later, Muslims to direct attention to Scripture than to the works of natural philosophy.

There is much one might linger over in the medieval period, the growth of alchemy, for example, or the universal belief in various forms of magic. These had their own quite well-defined forms of rationality in the service of practical goals. But the single episode to which I will restrict myself is the attack from the mid-thirteenth century onwards on the linked Aristotelian notions of nature and demonstration. Many Christian theologians saw these as limiting God's power and freedom. If natural science is really of the necessary, then how can the Creator be truly free or truly omnipotent?

The theology of divine omnipotence led, therefore, to the proposing of a new ontology, an ontology of individuals linked by similarities instead of by common natures. A new sort of science was needed for such a universe, one based on an induction that takes the form of generalization from resemblances and issues in probable knowledge only. The goals of explanation and secure knowledge would both have to be weakened in the new nominalist scheme, though not abandoned. It was not so much that the older goals had been discovered in practice to be unattainable as that their attainment was from the start ruled out on theological grounds.

9. The Scientific Revolution

The seventeenth was, of course, the crucial century for our quest. It was then that the elements of the rationality

of the natural sciences gradually began to come together
in something like the form in which we know them today.
What brought about these changes? There is no single
answer, but part of the answer, it seems fair to say, was the
gradual modification of the older goals, still seen as desirable, in the light of new successes and in the memory of
past failures. Three developments, in particular, should be
recalled.

First was the construction of a single science of astronomy which could be seen as predictive and explanatory at
once. The standard account of mathematical astronomy in
the late medieval period was that it did no more than
"save the appearances." The mathematical models it used
were regarded by most as devices useful for prediction but
in no way indicative of real processes and hence not explanatory. This separation between prediction and explanation had been seen by some, like Averroes, as troubling,
as something to be overcome.[54] After all, if an explanatory
science *did* in fact give us the truth of things, why would it
not serve to predict? Copernicus urged the *explanatory*
merits of his heliocentric model over that of his main
rival, Ptolemy; he could not claim an advantage on the
score of prediction. Galileo likewise showed how the Co-
pernican model gave a simpler explanatory account by
attributing a double motion to the earth rather than a
puzzling variety of motions to larger and very distant objects. But it was only when Newton created a dynamics for
planetary motions that explanation and prediction could
finally be brought together in a single structure. The ad-
vantages of this union were manifest once it was seen to
be possible. What had prevented it from happening as
early as Ptolemy or even Aristotle was only that no plausible way of combining the two in astronomy could then
be found. It was seen that a properly explanatory science
ought to make prediction possible but it took a very long
time before a coherent way of achieving this end could be
constructed. And even then, there was some question

about the merits of the mechanics as explanation, a point to which I will return.

But a second sort of relationship between prediction and explanation proved even more important. New instruments, like the telescope and the microscope, were beginning to reveal worlds that lay far beyond the range of ordinary experience. Could the sort of intuition of essence on which the older notion of science had depended reach out to these new worlds? Descartes attempted a variant of the older view, substituting mathematics for the syllogism and clear and distinct ideas for essences. He had some limited success in this way in mechanics but it was gradually borne in on him that this method could not yield satisfactory answers regarding the hidden causes responsible for the multiplicity of phenomena he was attempting to explain in such areas as optics and meteorology.[55]

The alternative was to try *hypothesis*—that is, to conjecture as to the cause and then test this conjecture in some way. The idea was not new, of course; the Aristotelians of the Renaissance had already explored it, and Bacon had advocated it openly. But the price was high. It seemed to entail giving up on the older goal of a definitive knowledge and settling for something much less. Descartes fought against this implication and tried to find a way to direct hypothesis to the older goal. But he was unsuccessful in this, and gradually expectations began to change. Huyghens gives clear expression to this change in an often-quoted passage:

> Whereas the geometers prove their propositions by fixed and incontestable principles, here [in optics] the principles are verified by the conclusions drawn from them, the nature of these things not allowing of this being done otherwise. It is always possible to attain thereby to a degree of probability which very often is scarcely less than complete proof. To wit, when things which have been demonstrated by the principles that have been assumed correspond perfectly to the phenomena which experiment has brought under observation, espe-

cially when there are a great number of them, and further, especially, when one can imagine and foresee new phenomena which one employs, and when one finds that therein the fact corresponds to our prediction. But if all these probable proofs are found [in my work], as it seems to me they are, this ought to be a very strong confirmation of the success of my inquiry.[56]

One could hardly find a better concise description of retroductive method in any modern writer. Huyghens wrote these lines in 1678, but they were not published until 1690, three years after the appearance of Newton's *Principia*. Newton disagreed with Huyghens on this as on many other matters. He thought the rationality appropriate to science to be an inductive one, and became in the course of his career more and more reluctant to allow hypothetical constructs into science proper, although he was willing to admit them at the preliminary stage of inquiry.

Newton's hesitations were shared by many, both on rationalist and on empiricist grounds.[57] What gradually turned the tide was the growth of theories in fields like chemistry and optics that could not possibly be represented as being purely inductive and yet were manifestly successful in terms of coherence and fertility. The notion of *theory* as an explanatory hypothesis appealing to underlying causal structures finally begin to make its way. The criteria Huyghens mentions are much broader than those of deductive or inductive rationality. They are precisely the criteria one would expect once it be admitted that the world really *does* contain underlying structures that are causally responsible for what we observe. But it is entirely contingent that it should do so. The world might not have been of that kind. Or these structures might not have been reachable by us, as Locke feared.

It was a *discovery* on our part that this is the kind of world we live in. In that sense, the history of science was involved. But once the discovery was made, one can proceed to the *justification* of fertility, coherence, consistency, and the rest as the rational criteria for theory-assessment on the grounds either of logic or of practice. Kepler, Descartes, Huyghens, and other seventeenth-century writers who suggested appropriate criteria for theory, did so on logical grounds and on the assumption of an underlying structure indirectly knowable by us. Later writers, like Whewell, adverted rather more to successful practice.

One final point in regard to *theory* as the principal vehicle of scientific explanation may be noted. One criterion of theory is empirical adequacy, its ability to account for the data in hand. The ancient separation between explanation and prediction was now ended, since empirical adequacy amounts to predictive accuracy in such contexts as planetary astronomy, so that prediction becomes a test of explanatory success, and hence part of the rationality serving the larger goal of explanation.

The acceptance of hypothesis as an appropriate instrument of understanding brought with it other changes too. For example, it immediately became necessary to contrive situations in which the consequences of hypotheses could be unambiguously tested, since the line of validation now ran from consequence back to hypothesis rather than from principle down to conclusion. Thus, *experiment* rapidly became the accepted way of relating theory to observation. This, in turn, compelled the use of *instruments,* since the human senses are neither sufficiently sharp to provide the needed precision, nor sufficiently wide-ranging to react to all the phenomena the scientist finds of interest. The senses began to lose their importance except as a means of registering instrumental readings. Instrument-aided measurement required the use of the language of mathematics and the setting aside, as secondary, of properties that could not easily be quantified.

The story is a familiar one. I recall it here only to underline that these shifts were all interconnected in intricate ways. They were made for reasons that can in large part be reconstructed, reasons that bear on the connected aims of explanatory success, predictive power, and the strongest possible warrant for claims made. Choices had to be made. Not only was there the traditional Aristotelian mode of rationality, but there were long-standing alternatives like alchemy, magic, the Hermetic philosophy, Pythagorean number mysticism. We see these latter as nonstandard, but there was no simple means of making that judgment then. And many of the principal protagonists of the new science saw no incongruity in following the rationalities of the alchemist or the Hermetic on occasion.[58]

The gradual abandonment of these alternatives over the course of that eventful century was not, as Feyerabend argues, a matter of superior propaganda. I think it can be shown where each of them failed, and failed first and foremost in their *own* terms. I think it can be shown further how they failed in terms of the broad goals of explanatory power and secure knowledge inherited from the Greeks. I do not say that these failures were obvious in all cases at the time. But I do say that one can understand the gradual abandonment of these alternatives in terms of fairly specific reasons, reasons that would have counted as reasons then as well as now. Social factors, no doubt, played a role, as they always do when deep-routed intellectual beliefs change. But the success of the new science in its own terms could not be gainsaid, and its goals were in plausible continuity with the older goals of natural philosophy.

10. Recent Developments

In the time between Newton and our own, two developments are especially worth noting. The application of the new sciences to the improvement of the human condition, as Bacon had put it in the *New Organon,* proved much

longer in coming than he had anticipated. It was only in
the nineteenth century that theory had advanced far
enough in sciences such as chemistry or thermodynamics
to admit of direct technological application. And it has
really only been subsequent to World War II that the ex-
plosive developments of new technologies such as those of
communication have drawn directly on advances in theory
as the mainspring in their own progress. Traditional dis-
tinctions between theory and practice, between science
and technology, have become blurred, and have in many
contexts been superseded by a different distinction between research and development.

The importance of this for my topic is that the ancient
goal of human betterment in fields like medicine, agriculture, engineering, has now in large part blended with the
goals of natural science. Once science became predictive,
it was only a matter of time until those predictions yielded
practical fruit. It is the predictive side of scientific explanation that ultimately allows technological advantage to
become one of the goals of science itself. And so, another
tributary has flowed into the main stream, one that, to
pursue the metaphor, has greatly accelerated the rate of
flow of that stream. There would no longer be any way of
separating these goals, of returning to a time when technology and theoretical science could be viewed as unrelated. Now we *know* they are.

One further development is associated with the two
great theoretical advances in physics of our century, relativity theory and more especially quantum theory. But its
beginnings go back to Newton's day. What constitutes an
adequate explanation in mechanics? Descartes proposed
that only contact action could properly explain motion.
Then along came gravitation, which to all appearances
seemed like action at a distance. Most of Newton's con-
temporaries had reservations about the status of the new
mechanics as *science* because they could not be persuaded
that he had *explained* the motions of the planets, even

though it was agreed by all that he could *predict* them. Even Newton himself was uneasy, and tried over and over to reformulate the notion of gravitation in a more acceptable way. He did not succeed, and what happened was not that an acceptable reformulation was discovered, but that people eventually got used to the new idea.

The same sort of thing has happened in our own century. Einstein's Special Theory of Relativity violated people's physical intuitions by making the notion of simultaneity at a distance a relative one, dependent on the state of motion of the person making the simultaneity claim. Proponents of the new quantum theory were forced to abandon such cherished Newtonian presuppositions as precise location and momentum of elementary entities. This in turn undermines determinism of the classical sort. Even causality seems threatened. The decay of a single radium atom is in principle unpredictable; not only can no current theory predict it, no future theory can either, if the implications of recent analyses of the consequences of Bell's theorem are to be trusted. Does this mean it is uncaused? I think not, but it obviously forces us to reconsider what we mean by causality. Einstein regarded the new theory as lacking, on grounds of intelligibility, just as Leibniz did Newton's. But Einstein's judgment seems to have been set aside, just as Leibniz's was.

What does this do to explanation as a goal of science? The "ideals of natural order," as Toulmin called them, which we learned from the seventeenth century appear to be as much under pressure today as the Aristotelian ones were in the seventeenth century. This does not directly affect most of science; in fields like biology, the traditional notions of explanation still appear to hold good. It is only where reactions at a level lower than that of the atom become important, in theoretical physics or in cosmology, for example, that the older ideals fail.

That they should fail, at levels so remote from our ordinary middle-sized experience where these ideals were first formed (with the aid, let it be said, of Newtonian science), should not really be surprising. It simply means that the intuitions and expectations we draw from familiar experience are of doubtful service when we penetrate to the level of the electron or the quark. But if this is so, can we really speak of *explanation* or of *understanding* at that level at all? Must we not simply be content to say, as defenders of Newtonian mechanics long ago had to say to critics: it *works?* If a formalism works, in the sense of predicting accurately, does it ultimately lead to a new level of understanding or does it only lead to an *illusion* of understanding?[59]

Arthur Fine argues that quantum mechanics cannot be supposed to provide understanding: it is, he says, "the blackest of black-box theories, a marvelous predictor but an incompetent explainer."[60] Heisenberg took a somewhat different line. At the quantum level, he suggested, we have to substitute what he called a "Pythagorean" strategy for the older "Democritean" one which sought to explain causally in terms of underlying structure.[61] Instead of postulating entities ever smaller in size, we would look for mathematical symmetries, leaving aside all physical analogies. The recent successes of the quark model rather counts against Heisenberg's proposal, since quarks can still be thought of as physical structures, though with properties of a largely unintuitive sort. But as research continues, might not Heisenberg turn out to be right? Would this be equivalent to abandoning the goal of explanation at the microlevel and contenting ourselves with prediction only, as Fine supposes we have already done?

Heisenberg did not think so, as his choice of the label 'Pythagorean' indicated. He thought that a new, properly *mathematical,* form of understanding, with its own norms, would succeed the older, more familiar, physical one. And the symmetry principles that guide recent work in theoretical physics would be thought by many to provide a distinctively "physical" type of understanding of physical process at the most basic level. But we simply cannot anticipate what the future will bring in this regard, any more than Newton could have anticipated Heisenberg's own Uncertainty Principle. It is possible that as we move further and further away from the categories and analogies of our middle-sized world, we might have to settle in physics for a pure formalism which gives us all we want in the way of prediction but cannot in any sense be construed as causal understanding. My own inclinations would lead me to doubt this, but there can be no security about prophecy in this area. It depends on the as yet unprobed nature of the world, and for that matter, on the capacities of mind.

11. Conclusion

What have we learned from this story? The rationality of the activity we call science has in the centuries since Ptolemy's day allowed us to predict—that is, to specify—with ever-increasing accuracy not only the paths of the planets but a myriad of intricate processes on earth, and in doing so it has gradually transformed the ancient crafts into the powerful technologies that are altering our lives so rapidly today. These are permanent achievements, not in the sense that they might not one day be forgotten nor in the sense that their effects have always been beneficial. But they do represent the realization of the linked goals of predictive accuracy and technical control. Insofar as scientific rationality is a means to the realization of goals, it has clearly succeeded to a degree that could never have been anticipated. No later changes of belief on our part about the nature of the world can undo this success. This is something on which realist and antirealist can agree.

What would it mean, then, for the goals to change gradually until they were quite different? At that point, presumably, predictive accuracy would no longer matter;

"scientists" would lose interest in technological applications. Scientific theories (if there were theories) would no longer be evaluated by their success in accounting for the data. Meanwhile, it would still be possible, presumably, to achieve the older goals in the manner in which this is done today. In the circumstances, which would *we* call science? The answer is obvious, but perhaps insufficient. The issue is not word use, but whether the achievements of natural science up to this point have some sort of permanence about them. Could the notion of science change at every step for "reasons" as Shapere prescribes, so that what we now count as science would at a later time no longer qualify?

The naturalist is uneasy about constraints of an apparently "absolute" sort; it sounds as though an appeal is being made to an *a priori*. But the source of constraint can lie instead in what has been learned. Nor is an essence or a natural kind involved, if these are understood to impose a strong demarcation between science and nonscience. Of course, a "kind" of a very loose sort has been established by two historical discoveries; one is that certain goals are achievable and the other is that certain means (experimental procedures, epistemic values, etc.) serve these ends. These goals and the rational procedures that support them are sufficiently distinctive to mark off science, in a rough way only but still quite definitively, from various other human activities. Once stated, the point seems an altogether banal one. But since it has so often been called in question of late, it seems to merit reiteration.

It *might,* of course, happen that at some later time a reliable predictive knowledge of Nature will cease to seem a worthwhile goal; science-fiction writers have invited us to imagine a time when, in the aftermath of a nuclear holocaust, all books would be burned.[62] I am inclined to think that the quest for a usable knowledge of Nature is, if not an invariant of human nature, at least something that can scarcely be given up entirely. But to say that the goals of science would no longer be pursued is not at all the same as to say that these goals would progressively alter into something quite different. If the complex engines of scientific rationality, as we know it, were to be entirely stilled, there would be no successor activity that could plausibly be titled science.

Empirical adequacy, predictive success, technological utility, are reasonably well-defined goals and the extent to which they have been attained can be asserted with some confidence. But those who argue for the variability in principle of scientific rationality rely on the premise that what counts as *explanation* has varied widely over the history of science, even within the same field of inquiry. To the extent that science be thought of as a search for *understanding,* then, one might infer that it could take very different forms in the future. The notion of understanding (unlike that of prediction) is sufficiently indefinite (it is argued) that it could be a light constraint to satisfy.

Yet there *is* a real constraint here on what may count as natural science in the future. It will have to provide an understanding of natural process, and will thus (unlike football!) have to be directed in some systematic way to natural process. What counts as explanation varies greatly from one part of natural science to another. But there is one common bond. What makes scientific knowledge ampliative, what finally enables it to transcend the limits of the here and now, is its successful employment of retroduction. The criteria of theoretical explanation are sufficiently well-determined to enable us to infer from effect to unobserved cause with a fair degree of confidence. Predictive accuracy reappears here as a means, rather than as an end. As we saw earlier, one test (but by no means a sufficient test) for a good explanation is that it should "account for the data," that it should predict correctly.

There are obvious connections between rationality-change and realism. Someone who maintains the in-principle unlimited variability of the goals and procedures constituting scientific rationality would be likely (so it would seem) to reject existence-claims for the explanatory entities of even the most successful current theory. If what now counts as genuine causal understanding might simply cease to qualify as understanding of any sort, without any change in the specific relationship between evidence and hypothesis, then existence claims for the entities postulated by even the best supported theories of today (e.g., molecules, genes, dinosaurs, galaxies) would have yet another challenge to face. Critics of realism, like Laudan and van Fraassen, argue that retroduction does not ever of itself warrant existence-claims for the entities postulated as causes; if they could add that retroduction itself might prove only a passing fad, it would surely strengthen their case. Without retroduction as the basic explanatory strategy, the issue of realism might not even arise.

The issue in both cases is to assess the implications of change. Earlier accounts of scientific theory suggested that it would over the course of time converge on the real, as it were. Historicist claims about theory-replacement, prompted especially by Kuhn's work, have led many to question whether any such convergence is occurring, whether there is any cumulative growth in our knowledge of the underlying structure of the world. Likewise, older accounts of scientific rationality assimilated it to a more or less unchanging logic. The challenges we have been tracing infer from changes in the notion of understanding to the contingent character of the present notion. The question here, as in the case of realism, is how much of the present will carry over into the future. How much have we learned? What degree of permanence can reasonably be attributed to certain theoretical structures (realism) or certain goals and procedures (rationality)? The connection, of course, is that it is the goals and procedures that have enabled the structures to be discovered, while it is the discovery of the structures that (in part, at least) assures us that the goals and procedures are appropriate.

How might the argument run? Two alternatives suggest themselves. One is to take realism to be a well-supported doctrine, not as a blanket claim for all theory but as an inference, in the case of some theories only, from explanatory success of a certain durable sort. What would then authenticate certain criteria as "rational" would be the long-term explanatory success of theories validated by means of those criteria. The question: Is simplicity a rational criterion of theory-truth, would then be answered by looking at the record of how well this criterion has served. How well did hypotheses recommended, to a significant extent, by their "simplicity" survive to become part of the larger background of accepted theory?

But the argument might take another and more controversial form. Why is fertility an appropriate criterion of theory? Because, if the world has an underlying causal structure, this is a feature one would *expect* theory to possess. Are we back to an *a priori* here? No, because we need the testimony of history to assure us that the world does indeed (at least over large domains) have an underlying causal structure. But once we know this (and this is where realism comes in), then the choice of the "rational" criteria for theory-assessment begins to sort itself out.

In this case, the warrant for calling a particular criterion or procedure "rational" would not simply be the *fact* of past success, in purely instrumental terms. It would be the broadly logical connection perceivable between means and end, given a certain general sort of structure in the domain being investigated. The constraint laid on the future would then proceed from a dual source: the realist claim about the sort of connectedness to expect at a particular level or in a particular domain, and the logical claim as to what procedures are likely to prove appropriate. The latter can (and should) be supplemented by reference to the outcome of historical practice.

Here, then, may be an asymmetry between the issues of realism and rationality.[63] Only empirical tests can warrant theoretical models in the natural sciences; logical resources will not serve to determine the appropriateness of a particular physical model. Whereas, at the second level, the claim that, for example, what Copernicus called "naturalness" is an appropriate value to seek for in a theory can be attested in quasi-logical terms. Quasi-logical, because some very general features of the domain to be explained have to be presupposed.

This latter style of argument would, of course, be unacceptable from the naturalist standpoint. The naturalist extends Hume's inductivism a step upwards: regular recurrence—in this case, continued "success" of a particular procedure—is to provide the only acceptable testimony. The question as to why this kind of regularity occurs in the first place, why the procedure is successful, is blocked. Hume did not permit (did not, it must be said, adequately consider) the use of retroduction to infer to unobserved causes. The naturalist in our debate does not allow that mind may come to possess at least a partial understanding of understanding.

The issues here have all of a sudden broadened. And the options turn out to be familiar ones. If I leave them unresolved, I hope that I have at least indicated some profitable lines to pursue further.

NOTES

1. Kuhn, *The Structure of Scientific Revolutions* (Chicago: University of Chicago Press, 2nd ed., 1970), p. 184. (Abbreviated as *SSR* below).

2. Ibid., p. 185.

3. Ibid.

4. Ibid., pp. 61, 92.

5. Kuhn, *The Essential Tension* (Chicago: University of Chicago Press, 1977), p. 335.

6. See, for example, *SSR* pp. 85, 94.

7. *SSR*, p. 148.

8. Ibid.

9. Scheffler, *Science and Subjectivity* (Indianapolis: Bobbs-Merrill,

1967), pp. 9–10. Cited by Larry Laudan, *Progress and Its Problems* (Berkeley: University of California Press, 1977), p. 129.

10. Zahar, "Feyerabend on Observation and Empirical Content," *British Journal for the Philosophy of Science* 33 (1982), 397–408; see p. 407.

11. Ibid.

12. Feyerabend, *Against Method* (London: New Left Bookstore, 1975), pp. 27–28.

13. Ibid., p. 23.

14. Feyerabend, *Science in a Free Society* (London: New Left Bookstore, 1978), p. 32.

15. *Against Method,* p. 205. Several of the phrases here (specifically "of the last two centuries") are being quoted from Imre Lakatos, "History of Science and Its Rational Reconstructions," *Boston Studies in the Philosophy of Science,* vol. 8 (Dordrecht: Reidel, 1971), 91–136; see p. 111.

16. *Against Method,* p. 208.

17. Ibid., pp. 208–9.

18. Ibid., p. 209.

19. *Science in a Free Society,* p. 38.

20. Ibid., p. 33.

21. Ibid., p. 34.

22. See McMullin, "Philosophy of Science and Its Rational Reconstructions," in *Progress and Rationality in Science,* ed. G. Radnitzky and G. Anderson (Dordrecht: Reidel, 1978), pp. 201–32.

23. The example is David Bloor's in *Knowledge and Social Imagery* (London: Routledge, 1976), p. 26.

24. Barry Barnes, *Scientific Knowledge and Sociological Theory* (London: Routledge, 1974), p. 64. For a perceptive critique, see Steven Lukes, "Relativism in Its Place," in *Rationality and Relativism,* ed. Martin Hollis and Steven Lukes (Cambridge, Mass.: MIT Press, 1982), pp. 261–305.

25. Barry Barnes, *Interests and the Growth of Knowledge* (London: Routledge, 1977), chap. 1.

26. Bloor, *Knowledge and Social Imagery,* p. 39.

27. Laudan, *Science and Values* (Berkeley: University of California Press, 1984), pp. 63–64.

28. Ibid., p. 64.

29. Ibid.

30. Ibid., p. 65.

31. Ibid., p. 66.

32. Laudan, *Progress and Its Problems* (Berkeley: University of California Press, 1977), p. 130.

33. Ibid., p. 124.

34. Ibid., 130–31.

35. Ibid., p. 132.

36. "I hold now, much as I did ten years ago, with the pragmatists that science is a problem-solving activity. (But, of course, so are many other activities). That is true of all times and places since, in my view, problem-solving is constitutive of human cognition about the world." Private communication from L. L. to the author, 9 July 1986.

37. Shapere, "The Character of Scientific Change," in *Scientific Discovery, Logic, and Rationality*, ed. Thomas Nickles (Dordrecht: Reidel, 1980), pp. 61–116; reprinted in D. Shapere, *Reason and the Search for Knowledge* (Dordrecht: Reidel, 1984), 205–60. See p. 207 in the latter version.

38. Shapere, *Reason and the Search for Knowledge*, p. 208.

39. Ibid., p. 212.

40. Ibid., p. 223.

41. Ibid., p. 239.

42. By Gary Gutting, Larry Laudan, Michael Gardner, Thomas Nickles, and Ernan McMullin in ibid., pp. 246–56.

43. Ibid., p. 246.

44. Ibid., p. 247.

45. Shapere, "Objectivity, Rationality, and Scientific Change," *PSA 1984*, vol. 2, ed. Peter Asquith and Philip Kitcher (East Lansing, Mich.: Philosophy of Science Association, 1985), 637–63; see pp. 650–51.

46. Ibid., p. 653.

47. Ibid.

48. Ibid., p. 655.

49. Shapere, "The Character of Scientific Change: Panel Discussion," *Reason and the Search for Knowledge*, p. 247.

50. For a fuller analysis of the notion of value in this context, see my "Values in Science," *PSA 1982*, vol. 2, ed. Peter Asquith and Thomas Nickles (East Lansing, Mich.: Philosophy of Science Association, 1983), 3–25.

51. G. E. R. Lloyd, *Magic, Reason and Experience* (Cambridge: Cambridge University Press, 1979). This book is a model of how the inquiry into the origins of scientific rationality should be pursued.

52. See my "Truth and Explanatory Success," *Proceedings American Catholic Philosophical Association* 59 (1985), 206–31.

53. I have discussed these issues in much more detail in "The Goals of Natural Science," *Proceedings American Philosophical Association* 58 (1984), 37–64.

54. See "The Goals of Natural Science," section 3.

55. For a detailed treatment, see my "Conceptions of Science in the Scientific Revolution," in *Reappraisals of the Scientific Revolution,* ed. D. Lindberg and R. Westman, forthcoming from Cambridge University Press.

56. Huyghens, *Treatise on Light,* trans. S. P. Thompson (London: Macmillan, 1912), pp. vi–vii; translation slightly modified.

57. Laudan, *Science and Hypothesis* (Dordrecht: Reidel, 1981), chap. 7.

58. For illustrations, see *Occult and Scientific Mentalities in the Renaissance,* ed. Brian Vickers (Cambridge: Cambridge University Press, 1984).

59. See Oliver Costa de Beauregard, *American Journal of Physics* 51 (1983), 515–16. I am indebted to Jim Cushing for drawing my attention to this and the following reference.

60. Fine, "Antinomies of Entanglement," *Journal of Philosophy* 79 (1982), 733–48; see p. 740

61. Werner Heisenberg, "Tradition in Science," in *The Nature of Scientific Discovery,* ed. Owen Gingerich (Washington: Smithsonian Press, 1975), pp. 219–36.

62. The classic of the genre is Walter M. Miller's *A Canticle for Leibowitz* (New York: Lippincott, 1959).

63. In this paragraph, I am once more joining issue with Shapere, whose notion of "internalization" would not allow the sort of asymmetry I suggest.

IS NATURAL SCIENCE A NATURAL KIND?

Richard Rorty

1. Introduction

One of the principal reasons for the development of a subarea within philosophy called "philosophy of science" was the belief that 'science' (or, at least, 'natural science') named a natural kind, an area of culture which could be demarcated by one or both of two features: a special method, or a special relation to reality. The further suggestion, implicit in Carnap's work and made explicit by Quine, that "philosophy of science is philosophy enough," was a natural extension of this belief. For just as Plato was content to leave the world of appearances to the philodoxers, so many of the logical empiricists were, implicitly or explicitly, content to leave the rest of culture to itself. On their view, once the job of demarcation had been accomplished, once the distinctive nature of science had been accurately described, there was no need to say much about the other activities of human beings. For, since man was a rational animal and science the acme of rationality, science was the *paradigmatic* human activity. What little there was to say about other areas of culture amounted to a wistful hope that some of them (e.g., philosophy) might themselves become more "scientific."[1]

Hempel and others, however, showed that demarcation was not as easy as it had first appeared. The increasing plausibility of Neurathian holism, once it had been revivi-

fied by Quine's "Two Dogmas" and by Wittgenstein's
Philosophical Investigations, further undermined attempts to
isolate "the scientific method," because they undermined
attempts to isolate piecemeal connections between scientific theories and the world. Some philosophers followed
Hempel in dropping both the question "how do we demarcate science from metaphysics?" and metaphysics itself. These philosophers turned to attempts to construct a
logic of confirmation, without worrying greatly about
whether the use of such logic distinguished science from
nonscience. But other philosophers followed Quine in falling back into dogmatic metaphysics, decreeing that the
vocabulary of the physical sciences "limns the true and
ultimate structure of reality." It is significant that Quine
concluded that "the unit of empirical inquiry is the whole
of science," when one might have expected, given the drift
of his argument, "the whole of culture." Quine, and many
other holists, persisted in the belief that the science-
nonscience distinction somehow cuts culture at a philosophically significant joint.

The cash value of the claim that it does so is a refusal to
rest content with a merely Baconian criterion for separating science from nonscience. On the (familiar, if Whiggish) interpretation of Bacon common to Macaulay and
Dewey, Baconians will call a cultural achievement "science" only if they can trace some technological advance,
some increase in our ability to predict and control, back
to that development. (That is why Baconians boggle at the
phrase 'Aristotelian science'.)

This pragmatic view that science is whatever gives us
this particular sort of power will be welcome if one has
developed doubts about traditional philosophical inquiries into scientific method and into the relation of science
to reality. For it lets us avoid conundrums like "what
method is common to paleontology and particle physics?"
or "what relation to reality is shared by topology and ento-
mology?" while still explaining why we use the word 'science' to cover all four disciplines. The same view lets us

treat questions like "is sociology a science?" (or, "can the social sciences be as scientific as the natural sciences?") as empirical (indeed, sociological), questions about the uses to which the work of social scientists has been or might be put. This Baconian way of defining 'science' is, of course, no less fuzzy than the notion, *prediction* and *control.* Despite this fuzziness, it is probably the one most frequently employed by deans, bureaucrats, philanthropoids, and the lay public.

Since the forties, the period when Hempel and Quine began questioning the logical empiricists' basic assumptions, there have been two further stages in the discussion of the question of whether science is a natural kind. The first stage concentrated on the notion of method and centered around the work of Kuhn and Feyerabend. The second, in the midst of which we now find ourselves, concentrates on the question of science's relation to reality, and revolves around the ambiguous term 'scientific realism'.

The fracas over Kuhn's and Feyerabend's claim that some scientific theories were incommensurable with predecessor theories was created by philosophers who were intent upon salvaging a nonpragmatic criterion for distinguishing science from nonscience. Most of Kuhn's readers were prepared to admit that there were areas of culture— e.g., art and politics—in which vocabularies, discourses, Foucaultian "*epistēmēs*" replaced one another, and to grant that, in these areas, there was no overarching metavocabulary into which every such vocabulary might be translated. But the suggestion that this was true of the natural sciences as well was found offensive. Critics of Kuhn such as Scheffler and Newton-Smith thought of Kuhn as casting doubt on "the rationality of science." They sympathized with Lakatos' description of Kuhn as having reduced science to "mob psychology."

But although these critics might have hesitated to say explicitly that politics and art were matters of "mob psychology," their position implied just that. Defenders of the idea that there is a methodological difference between artistic, political, and scientific revolutions typically adopt a strong, criterial, notion of rationality, one in which rationality is a matter of abiding by explicit principles. They thus find themselves, willy-nilly, questioning the "rationality" of the rest of culture. Kuhn's defenders, by contrast, typically draw the line between the rational and the nonrational sociologically (in terms of a distinction between persuasion and force) rather than methodologically (in terms of the distinction between possession and lack of explicit criteria).

The strong point of Kuhn's critics was that incommensurability seemed to entail indiscussibility. The strong point of his defenders was that, given Hempel's critique of verificationism and Quine's of the fact-language distinction, nobody could answer Kuhn's challenge by explaining how commensuration was possible. So the Kuhnian wars dragged on, with both sides talking past each other.

These wars now seem to be drawing to a close. For both sides are coming to agree that untranslatability does not entail unlearnability, and that learnability is all that is required to make discussability possible. Most of Kuhn's critics have conceded that there is no ahistorical meta-vocabulary in terms of which to formulate algorithms for theory-choice (algorithms which might actually be useful to practicing scientists, rather than being *post factum* constructs). Most of his defenders have conceded that the old and the new theories are all "about the same world." So there is little left for them to quarrel about. The effect of this reconciliation is that the attempt to avoid a merely pragmatic and Baconian definition of the term 'science' has swung away from the question of science's rationality toward that of its relation to the world— from method to metaphysics. The resulting shift of attention has caused discussion to center around three different topics, all of which are discussed under the heading of 'scientific realism'.

First, there is the topic of "different worlds." This topic is still on the table because there are still recalcitrant Kuhnians who take the claim that Aristotle and Galileo "lived in different worlds" literally. These diehards play into the hands of diehard adherents of Putnam's erstwhile view that only a causal theory of reference can save us from relativism. The two sets of diehards engage in what Arthur Fine has called "a fine metaphysical *pas de deux.*" Second, there is the topic of instrumentalism—of whether electrons *really exist,* in some sense of "really" in which tables uncontroversially do so exist. The distinction between "belief in x" and "heuristic use of the concept of x," discarded as verbal and "making no difference" by Deweyans such as Ernest Nagel and Sidney Morgenbesser, has recently been given a new lease on life by Michael Dummett, Bas van Fraassen, and others. Third, there is the claim, spelled out boldly and clearly by Bernard Williams, that science is distinguished from nonscience by the fact that although nonscience—e.g., art and politics—may, *pace* Plato, attain the status of "knowledge" and may converge to lasting agreement, nevertheless it differs from science in not being "guided" to such agreement by the way the world is in itself.

Let me dub the first topic—the one about many worlds—"realism versus relativism." I shall call the second issue, the one resurrected by van Fraassen, "realism versus instrumentalism," and the third, the one discussed by Williams, "realism versus pragmatism." As Ernan McMullin has noted, the term 'antirealism' covers too much ground. One has to be careful to keep distinct various positions which people who call themselves "realists" dislike. One must also note, with Fine, that Dummett's term 'antirealism' tends to beg the question which the pragmatist wants to raise: the question of whether notions like "made true by the world," "fact of the matter," and "ontological status" should be used at all, or rather discarded. Fine, for example, wants to find a position which is beyond realism and antirealism.[2]

Since I share this aim with Fine,[3] most of my paper will be devoted to the quarrel between realism and pragmatism, viewed as a quarrel about whether the notions in terms of which Williams states his brand of realism are useful ones. I see this as a reflection of the deeper quarrel about whether we should persist in trying to view science as a natural kind, instead of just falling back on the Baconian-Deweyan view of the matter. I want to defend the latter view by urging that three notions which are used to defend opposing views are very dubious. These are: (1) the notion of "the world making sentences true," a notion essential to the diehard Kuhnian's claim about "many worlds"; (2) the notion of "the abductive method," a notion essential to the quarrel between realism and instrumentalism; (3) Williams' notion of the world "guiding" the work of scientists and causing their opinions to converge. I shall treat the first two notions relatively briefly and dogmatically, and then concentrate on the third.

2. Realism versus Relativism

In order to see the role of the first of these notions, that of the world making beliefs true, consider the following inference:

(1) There is no way to *translate* the relevant portions of Aristotle's vocabulary into the relevant portions of Galileo's, although each could *learn* the other's vocabulary.

(2) So there is no way to argue against Aristotle's views on the basis of beliefs phrased in Galileo's vocabulary, nor conversely.

(3) So both Aristotle's and Galileo's views must be held to be true, and therefore the application of the term 'true' must be relativized to vocabularies.

(4) The world makes beliefs true.

(5) But the same world cannot make both Aristotle and Galileo true, and so *different* worlds must do so.

One can attack the inference to (5) in two ways: by questioning the step from (2) to (3) or by denying (4). I should want to do both, on the basis of the Davidsonian doctrine that 'true' does not name a relation between discourse and the world, and more generally that the term 'true' should not be analyzed or defined.[4] I would like to combine this doctrine with what I have elsewhere called "ethnocentrism,"[5] the view that our own present beliefs are the ones we use to decide how to apply the term 'true', even though 'true' cannot be *defined* in terms of those beliefs. Then we can admit (2) but deny (3) by saying that the internal coherence of either Aristotle or Galileo does not entitle their views to the term 'true', since only coherence with *our* views could do that. We thus get a position which *trivializes* the term 'true' (by detaching it from what Putnam calls a "God's-eye view") but does not *relativize* it (by defining it in terms of some specific "conceptual scheme").[6]

One consequence of this position is that we should not think of the relation between inquiry and the world on what Davidson calls the "scheme-content" model. Another is that, as Davidson says:

> all the evidence there is is just what it takes to make our sentences or theories true. Nothing, however, no thing, makes sentences or theories true: not experience, not surface irritations, not the world, can make a sentence true.[7]

In other words, the equivalences between the two sides of Tarskian T-sentences do not parallel causal relationships which link sentences to nonsentences. This denial of (4) is, for purposes of arguing against relativism, more important than the denial of (3). For it gets at the essential point that there is no way to divide up the true sentences into those which express "matter of fact" and those which do not, and *a fortiori* no way to divide them up into those which express facts about one world and those which express facts about another.

Trivializing 'true' in the way Davidson does—holding that the reason this term is not synonymous with "justified by our lights" is not that it is synonymous with "justified by the world's lights" but because it is not synonymous with *anything*—seems to me the best way to *aufheben* both the diehard Kuhnian "different world" thesis and the diehard Putnamesque claim that only a nonintentional theory of reference can save us from relativism.[8] On this Davidsonian view, every sentence anybody has ever used will refer to the world *we* now believe to exist (e.g., the world of electrons and such). This claim, however, is not—as it once was for Putnam—the controversial result of a new, Kripkean, theory of reference. It is as trivial as the claim that Aristotle and Galileo both have to face the tribunal of our present beliefs before we shall call anything either said "true."

This is all that I want to say about realism versus relativism. As far as I can see, relativism (either in the form of "many truths" or "many worlds") could only enter the mind of somebody who, like Plato and Dummett, was antecedently convinced that some of our true beliefs are related to the world in a way in which others are not. So I am inclined to think that Kuhn himself was unconsciously attached to such a distinction, despite the fact that *The Structure of Scientific Revolutions* has done so much to undermine the Platonic distinction between *epistēmē* and *doxa*. If one drops that distinction and follows through on Quinean holism, one will not try to mark off "the whole of science" from "the whole of culture," but instead will see all our beliefs and desires as part of the same Quinean web. The web will not, *pace* Quine, divide into the bit which limns the true structure of reality and the part which does not. For carrying through on Quine brings one to Davidson: to the refusal to see either mind or languages as standing to the rest of the world as scheme to content.

3. Realism versus Instrumentalism

Let me try to be equally brief on the subject of realism versus instrumentalism. I do not want to take up the question of whether we can get an interesting distinction between the observable and the unobservable. Rather, I want to focus on some questions about the relations between pragmatism and instrumentalism which have been raised by McMullin. McMullin, commenting on the work of Putnam and myself, writes as follows:

> Recall that the original motivation for the doctrine of scientific realism was not a perverse philosopher's desire to inquire into the unknowable or to show that only the scientist's entities are "really real." It was a response to the challenges of fictionalism and instrumentalism, which over and over again in the history of science asserted that the entities of the scientist are fictional, that they do not exist in the everyday sense in which chairs and goldfish do. Now, how does Rorty respond to this? Has he an argument to offer? If he has, it would be an argument for scientific realism. It would also (as far as I can see) be a return to philosophy in the "old style" that he thinks we ought to have outgrown.[9]

My answer to McMullin's question is that we pragmatists try to distinguish ourselves from instrumentalists not by arguing against their answers but against their questions. Unless one were worried about the really real, unless one had already bought in on Plato's claim that degrees of certainty, or of centrality to our belief system, were correlated with different relations to reality, one would not know what was meant by "the everyday sense of existence." It takes, after all, a good deal of acculturation to get the point of questions like "Do numbers, or justice, or God, exist in the sense that goldfish do?" Before we can get our students to approach these questions with appropriate respect, we have to inculcate a specifically philosophical use of the term 'existence', one in which it is pretty well interchangeable with 'ontological status'. I do not think that this use can be taught unless the teacher at least hints at an invidious hierarchy—the divided line, the primary-versus-secondary quality distinction, the distinction between canonical and noncanonical notations, or something of the sort. We pragmatists think that once we stop taking such hierarchies seriously we shall see instrumentalism as just a quaint form of late Platonism.

So I think the only argument we pragmatists need against the instrumentalist is the one McMullin himself gives when he says "The realist claim is that the scientist is discovering the structures of the world; it is not required that these structures be imaginable in the categories of the macroworld."[10] But this is not a return to philosophy in the old style, nor is it really an "argument." It is just an attempt to shift the burden of argument to the instrumentalist by asking him: why do you attach more importance to the features which goldfish have and electrons lack than to the features which goldfish have and tables lack?

One popular instrumentalist answer to this question is: "because I am an empiricist." But this just shoves the issue back a step. Why, we Davidsonians want to know, does the instrumentalist think that some beliefs (e.g., about goldfish) are made true by experience? This question can be broken down into: (1) Why does he think they are made true by anything? (2) Why does he think that experience—in the sense of "the product of the human sense-organs"—has a crucial role to play with respect to certain beliefs and not to others? I shall defer the former, more general, question until my discussion of Williams. But let me venture a quick and partial answer to the second, viz.: the instrumentalist thinks this because he thinks that there is a special method, peculiarly bound up with modern science, called 'abduction', whose results stand in contrast to "the evidence of the senses."

Many of the philosophers of science whom I most admire, including McMullin, Sellars, and Fine, are guilty of encouraging the instrumentalist in this belief. McMullin, for example, begins the article which I have been citing by saying that

> When Galileo argued that the familiar patterns of light and shade on the face of the full moon could best be accounted for by supposing the moon to possess mountains and seas like those of earth, he was employing a joint mode of inference and explanation that was by no means new to natural science but which since then has come to be recognized as central to scientific explanation.[11]

Here McMullin lends aid and comfort to the idea that "scientific explanation" is explanation of a distinctive sort—that science can be distinguished from nonscience by its use of a special sort of inference. One suspects that he might agree with Clark Glymour that the principal motive of philosophy of science is to provide what Glymour calls: "a plausible and precise theory of scientific reasoning and argument: a theory that will abstract general patterns from the concreta of debates over genes and spectra and fields and delinquency."[12]

Fine too sometimes writes as if we all knew what a certain sort of inference called "abductive" is. Boyd represents Fine's argument against scientific realism accurately, I think, when he says that Fine accuses the realist of using an abductive argument for the nature of reality when explaining the success of science. Boyd concludes that Fine thus begs the question against the instrumentalist. For the instrumentalist has doubts about whether, in Boyd's words, "abduction is an epistemologically justifiable inferential principle, especially when, as in the present case, the explanation postulated involves the operation of unobservable mechanisms."[13]

It seems safe to say that almost everybody who tries to resolve, rather than dissolve, the issue of realism vs. instrumentalism takes for granted that we can find something like an "inferential principle" which can be called "abductive" and which is more prevalent in modern science than in, say, Homeric theology or transcendental philosophy. My own, strictly amateurish, guess would be that any "inferential principle" which is "central to scientific explanation" is going to turn out to be central to practically every other area of culture. In particular, postulating things you can't see to explain things you can see seems no more specific to those activities normally called "science" than is *modus ponens*. The last fifty years' worth of attempts to give Glymour what he wants suggests that we shall find nothing which both meets Glymour's requirements and is specific to what has traditionally been called "science."

4. Realism versus Pragmatism

With this dogmatic claim, I pass on from the issue of realism versus instrumentalism to my principal topic—realism versus pragmatism. Not only does the absence of an inferential principle specific to science make it hard for the instrumentalist to answer questions about why the observable-unobservable distinction matters, it also makes it hard for the realist who wants to claim that realism "explains the success of science." The reason is, once again, that the absence of a way of isolating a specifically scientific method makes the nature of the *explanandum* unclear. For realists badly need the idea that "science" is a natural kind.

It is not enough for them, e.g., to explain the success of technology based on belief in elementary particles by the existence of elementary particles. For they recognize that this sort of explanation is trivial. All it does is to say that we describe our successful actions as we do because we hold the theories we hold. Such an explanation of current success is as vacuous as our ancestors' explanations of past successes. (Why are we able to predict eclipses so well? Because Ptolemy's *Almagest* is an accurate representation of the heavens. Why is Islam so spectacularly successful? Because of the will of Allah. Why is a third of the world Communist? Because history really *is* the history of class struggle.)

To get beyond such vacuity, the realist must explain something called "science" on the basis of something called "the relation of scientific inquiry to reality"—a relation not possessed by all other human activities. So, to get his project off the ground, he must have in hand some independent criterion of scientificity other than this relation to reality. He wants to claim that "because there really *are* elementary particles" is part of the best explanation of the success of IBM, whereas "because history really *is* the history of class struggle" is no part of the best explanation of the success of the KGB. So he must find some feature of elementary particle theory which makes it an example of "science" and does not also make Marxist theory "science." It is hard to see how this feature could be other than a methodological one.

This point is made by Boyd—who seems to me not to realize how hot the water is into which he is prepared to plunge his fellow realists. Boyd says that:

> When philosophers of whatever persuasion assert that the methods of science are instrumentally (or theoretically, for that matter) reliable, their claim is of very little interest if nothing can be said about which methods are the methods in question. . . . Moreover, it will not do to countenance as "methods of science" just any regularities that may be discerned in the practice of scientists. If the reliability thesis is to be correctly formulated, one must identify those features of scientific practice that contribute to its instrumental reliability.[14]

This point complements a point made by Michael Levin, who notes that any realist who wants to explain the success of a scientific theory by reference to its truth had better answer the question "what kind of *mechanism* is truth?"[15] If realists are going to do any explaining that is not of the "dormitive power" sort they are going to have to describe two bits of mechanism and show how they interlock. They are going to have to isolate some reliability-inducing methods which are not shared with all the rest of culture and then isolate some features of the world which gear in with these methods. They need, so to speak, two independently describable sets of cogwheels, exhibited in sufficiently fine detail so that we can see just how they mesh.

To illustrate how far contemporary discussion of the realism versus pragmatism issue is from any attempt to offer such detailed descriptions, consider Bernard Williams' defense of his claim that: "in scientific [as opposed to ethical] inquiry there should ideally be convergence on an answer, where the best explanation of the convergence involves the idea that the answer represents how things are."[16]

Williams offers a reply to the Davidsonian objection that notions like "how things are" or "the world" (and, *a fortiori,* truth defined as "correspondence to the world") cannot explain anything because each of these is "an empty notion of something completely unspecified and unspecifiable."[17] His reply consists in the suggestion that we can form the idea of "an absolute conception of reality" as one which "might be arrived at by any investigators, even if they were very different from us."[18] Our present scientific theories, he thinks, tell us that "*green,* for certain, and probably *grass* are concepts that would not be available to every competent observer of the world and would not figure in the absolute conception." He continues:

> The substance of the absolute conception (as opposed to those vacuous or vanishing ideas of "the world" that were offered before) lies in the idea that it could non-vacuously

explain how it itself, and the various perspectival views of the world, are possible.[19]

To explain how a set of beliefs *is possible* is a high transcendental task, one which contrasts with simply explaining why these beliefs rather than others are actual. The latter sort of explanation is provided by intellectual history, including the history of science. That sort of explanation is not good enough for Williams. For he thinks it stays on a "perspectival" level, the level of beliefs and desires succeeding one another and interacting with one another over the course of time. Such an explanation of convergence of belief cannot, in Williams' view, be "the best explanation." The best explanation, presumably, would be one which gives us the sort of thing Levin wants, a mechanistic account as opposed to one phrased in intentional terms. It would be one which, in Williams' terms, shows how "convergence has been guided by the way things actually are,"[20]—one which spells out the details of this "guiding" in a way in which, e.g., a theological explanation of the success of Islam cannot spell out the operation of the will of Allah.

Williams does not try to spell out these details, but instead relies on the claim that such a spelling out is in principle possible and that, when actual, it would constitute the "best explanation" of the success of science. His approach to the science-ethics distinction, and his deprecation of explanations by reference to belief and desire, parallels a line taken by Gilbert Harman and Thomas Nagel. Consider the following passage from Harman, quoted approvingly by Nagel:

> Observation plays a role in science that it does not seem to play in ethics. The difference is that you need to make assumptions about certain physical facts to explain the occurrence of the observations that support a scientific theory, but you do not seem to need to make assumptions about any moral facts to explain the occurrence of . . . so-called moral observations. . . . In the moral case, it would seem that you

need only make assumptions about the psychology or moral sensibility of the person making the moral observation.[21]

It seems to me that we can explain the observations made—that is, the beliefs acquired without inference—by *both* the moralist and the scientist by reference simply to their respective "psychologies and sensibilities." For both, we can explain propensities to react with certain sentences to certain stimuli—stimuli described in neutral psychologese—by reference to their upbringing. The scientists have been programmed so as to respond to certain retinal patterns with "there goes a neutrino," just as the moralists have been programmed to respond to others with "that's morally despicable." One would naturally assume that the explanation of how a given human organism got programmed to make noninferential reports in a given vocabulary would contain about nine parts intellectual history to one part psycho-physiology. This would seem as true for scientists as for moralists.

But for Williams, Harman, and Nagel, such an explanation would not be the "best." The best explanation would be one which somehow replaced the intellectual history parts and used none but "nonperspectival" terms. Presumably one advantage these philosophers see in such a replacement, one criterion for "best" which they tacitly employ, is that such an explanation will tell us, as intellectual history does not, how the world causes us to acquire the vocabularies we employ. Further, it would provide what Mary Hesse has suggested (rightly, on my view) we are not going to get: a sense of "convergence" which covers convergence of concepts as well as of beliefs.[22] The history of science tells us only that one day Newton had a bright idea, namely *gravity*, but stays silent on how gravity caused Newton to acquire the concept of itself—or, more generally, how the world "guides" us to converge on "absolute" rather than merely "perspectival" terms. The best explanation will presumably fill this gap. It will do for *gravity, atom, quantum,* etc., what (supposedly) psycho-physiology does for *green*—explain how the universe, under a nonperspectival description, gets itself described both under that description and under perspectival ones.

But it is not clear that science can do this even for *green*, much less for *gravity*. Remember that what is in question is the first acquisition of a concept by a human being, not its transmission from the old to the young. Do we have the slightest idea of what happened when somebody first used some word roughly co-extensive with 'green'? Do we even know what we are looking for when we ask for an explanation of the addition of a concept to a repertoire of concepts, or of a metaphor to a language? Once we give up the Myth of the Given, the Lockean idea that (as Geach put it) when we invent "green" we are just translating from mental into English, there seems nowhere to turn.

The closest I can come to imagining what such an explanation would be like would be to describe what happens in the brain of the genius who suddenly uses new vocables, or old ones in new ways, thereby making possible what Mary Hesse calls "a metaphoric description of the domain of the explanandum."[23] Suppose that the psycho-physiology of the future tells us that the brains of linguistic innovators are hit by neutrinos at the right time in the right way. When the brains of certain language-using organisms are hit by neutrinos under certain conditions, these organisms blurt out sentences containing either neologisms like 'green' or metaphors like 'grace' or 'gravity'. Some of these neologisms and metaphors may then get picked up and bandied about by the organism's linguistic peers. Of these, those which "fit" either the world as it is in itself or our peculiarly human needs (the "nonperspectival" and the "perspectival" ones, respectively) will survive. They will be literalized and take their place in the language.

Given such an explanation, all we need is a way of telling which of these new concepts are perspectival and which absolute. We do so by reminding ourselves which we need to describe the acquisition of concepts. Sure enough: *neutrino* is on the list and *green* is not, just as Williams suspected. But this little fantasy takes us around a rather tiny circle. We had to know in advance in what sort of discourse an explanation of concept-formation might be formulated. Given our present scientific theories, the best we could think of was the discourse of neurophysiology. So we knew in advance that neither greenness nor Divine Grace nor the class struggle would turn up in the explanation of our acquisition of the terms 'green' or 'grace' or 'class struggle'. This is not an empirical discovery about how the world guided us. It is just physicalism employed as a regulative idea, a consequence of our present guesses about how we might some day explain something which we actually have no idea how to explain. But why should we say that a terminology which might conceivably enable us to do something we presently have no idea how to do is the best candidate for the "absolute" conception of reality?

This last question raises a more basic one: what is so special about prediction and control? Why should we think that explanations offered for this purpose are the "best" explanations? Why should we think that the tools which make possible the attainment of these particular human purposes are less "merely" human than those which make possible the attainment of beauty or justice? What is the relation between facilitating prediction and control and being "nonperspectival" or "mind-independent"?

For us pragmatists, the trail of the human serpent is, as William James said, over all. Williams finds it "obvious" that the pragmatist is wrong, that there is a difference between practical deliberation and the search for truth[24]—precisely the distinction which James tried to collapse when he said that "the true is the good in the way of belief." Yet even if we grant this dubious distinction for the sake of argument, we shall still want to know what special connection exists between the search for "nonperspectival" truth and the quest for beliefs which enable us to predict and control. As far as I can see, Williams also takes it as "obvious" that there is such a connection.

The argumentative impasse between Williams' realism and pragmatism is evident in the penultimate paragraph of Williams' paper "The Truth in Relativism":

> Phlogiston theory is, I take it, not now a real option; but I doubt that that just means that to try to live the life of a convinced phlogiston theorist in the contemporary Royal Society is as incoherent an enterprise as to try to live the life of a Teutonic knight in 1930's Nuremberg. One reason that phlogiston theory is not a real option is that it cannot be squared with a lot of what we know to be true.[25]

But from the pragmatist's point of view, that is just what the claim that phlogiston theory is not now a real option *does* mean. The two enterprises are on a par. Nowadays the beliefs essential to living the life of a Teutonic knight cannot be squared with what we know to be true. To see the analogy, all one needs is the same self-confidence in one's moral knowledge as the Royal Society has in its chemical knowledge. To prevent moral dogmatism, all one needs is the same open-mindedness which—one trusts—would permit the Royal Society to reinvent phlogiston if that happened to be what the next scientific revolution demanded.

In *Ethics and the Limits of Philosophy*, Williams changes his position slightly. He is now prepared to apply the honorific term 'knowledge' to the ethical beliefs most of us share, and from which would-be Teutonic knights dissent. But he still wants to keep a sharp line between science and ethics by claiming that, in the terms he used in his earlier article, "questions of appraisal do not genuinely arise" in the extreme cases of ethical disagreement, i.e., clashes of cultures. In his later terminology, this appears as the claim that "the disquotation principle"[26] applies to

the former but not to the latter. That is, he wants to relativize ethics, but not science, to membership in a culture.
The pragmatist wants to derelativize both by affirming
that in both we aim at what Williams thinks of as "absolute" truth, while denying that this latter notion can be
explicated in terms of the notion of "how things really
are." The pragmatist does not want to explicate 'true' at
all, and sees no point either in the absolute-relative distinction, or in the question of whether questions of appraisal *genuinely* arise. Unlike Williams, the pragmatist
sees *no* truth in relativism.

As far as I can see, all the apparent ways out of this
impasse just lead to other impasses. One could, for example, go back to the question of the perspectival character of *green* and run through the usual arguments
concerning Berkeley's claim that *mass* is equally perspectival.[27] But this would just wind up with an impasse at the
question of whether or not all words of a human language
are equally tainted with relativity to human interests. One
might try to break that impasse by asking about the correctness of Wittgenstein's picture of the relation of language to the world (one which allows no room for Thomas
Nagel's "subjective-objective" distinction or for Williams'
"genuine appraisal–nongenuine appraisal" distinction).
Nagel rightly says that Wittgenstein's view of how thought
is possible "clearly implies that any thought we can have
of a mind-independent reality must remain within the
boundaries set by our human form of life."[28] He concludes, following Kripke, that realism "cannot be reconciled with Wittgenstein's picture of language."

Wittgenstein's picture of the relation of language to the
world is much the same as Davidson's. They both want us
to see the relation as merely causal, rather than also representational. Both philosophers would like us to stop thinking that there is something called "language" which is a
"scheme" which can organize, or fit, or stand in some
other noncausal relation to, a "content" called "the

world."[29] So to discuss whether to give up Wittgenstein or to give up realism would be to bring us back around to the question of whether notions like "best explanation" can be employed *sans phrase.*

From a Wittgensteinian or Davidsonian or Deweyan angle, there is no such thing as "the best explanation" of anything; there is just the explanation which best suits the purpose of some given explainer. Explanation is, as Davidson says, always under a description, and alternative descriptions of the same causal process are useful for different purposes. There is no description which is somehow "closer" to the causal transactions being explained than the others. But the only sort of person who would be willing to take this relaxed pragmatic attitude toward alternative explanations would be somebody who was content to demarcate science in a merely Baconian way. So there seems little point in pursuing the issue between realism and pragmatism by switching from philosophy of science to philosophy of language. The impasses one comes to in either area look pretty much the same.

5. Scientificity as Moral Virtue

There is another way of breaking this impasse, but it is one which looks much more attractive to pragmatists than to realists. It is to ask the intellectual historian for an account of why the science versus nonscience distinction ever attained the importance it did. Why was there a demarcation problem in the first place? How did we ever start going around these circles?

One familiar attempt to answer this question starts with a claim which Williams discusses and dislikes: the thesis, common to Nietzsche and Dewey, that the attempt to distinguish practical deliberation from an impersonal and nonperspectival search for truth (the sort of search of which natural science is thought to be paradigmatic) is an attempt at "metaphysical comfort," the sort of comfort which was once provided by religion. Williams thinks that any such answer in terms of social psychology is "not in the least interesting."[30] He thereby adds one more disagreement to the list of those which divide realists and pragmatists. We pragmatists, following up on Hegel and Dewey, are very much interested in finding psycho-historical accounts of philosophical impasses. We particularly enjoy reading and writing dramatic narratives which describe how philosophers have backed themselves into the sort of corner which we take contemporary realists to be in.[31] For we hope that such narratives will serve therapeutic purposes, that they will make people so discouraged with certain issues that they will gradually drop the vocabulary in which those issues are formulated. For realists like Williams, on the other hand, this strategy is a sneaky way of avoiding the real issues—namely, issues about which explanations are best, best *sans phrase.*

Though I disagree with Williams about whether these issues are real, I agree with him that we should not be content to dismiss the idealization of science, the attempt to demarcate and then sacralize it, as *merely* an attempt at metaphysical comfort. For a second, complementary, psycho-historical answer to the question about the origin of the demarcation problem is available—an answer which can be made considerably more concrete and detailed, and one with which Williams might have some sympathy. This is that natural scientists have frequently been conspicuous exemplars of certain moral virtues. Scientists are deservedly famous for sticking to persuasion rather than force, for (relative) incorruptibility, for patience and reasonableness. The Royal Society and the circle of *libertins érudits* brought together, in the seventeenth century, a morally better class of people than those who were at home in the Oxford or the Sorbonne of the time. Even today, more honest, reliable, fair-minded people get elected to the Royal Society than to, for example, the House of Commons. In America, the National Academy of Sciences is notably less corruptible than the House of Representatives.

It is tempting—though, on a pragmatist view, illusory—to think that the prevalence of such virtues among scientists has something to do with the nature of their subject or of their procedures. In particular, the rhetoric of nineteenth-century scientism—of a period in which a new clerisy (typified by T. H. Huxley, as its predecessor was typified by his episcopal opponent) was coming to self-consciousness and developing a vocabulary of self-congratulation—confused these moral virtues with an intellectual virtue called "rationality." The attempt to find a non-Baconian way of demarcating science thus gets part of its justification from the assumption that we need a metaphysical (or, better yet, a physicalist) account of the relation between human faculties and the rest of the world, an account in which "reason" is the name of the crucial link between humanity and the nonhuman, our access to an "absolute conception of reality," the means by which the world "guides" us to a correct description of itself.

But if, as I do, one views pragmatism as a successor movement to romanticism, one will see this notion of reason as one of its principal targets.[32] So we pragmatists are inclined to say that that there is no *deep* explanation of why the same people who are good at providing us with technology also serve as good examples of certain moral virtues. That is just a historical accident, as is the fact that, in contemporary Russia and Poland, poets and novelists serve as the best examples of certain other moral virtues. On a pragmatist view, rationality is not the exercise of a faculty called "reason"—a faculty which stands in some determinate relation to reality. Nor is it the use of a method. It is *simply* a matter of being open and curious, and of relying on persuasion rather than force.

"Scientific rationality" is, on this view, a pleonasm, not a specification of a particular, and paradigmatic, kind of rationality, one whose nature might be clarified by a disci-

pline called "philosophy of science." We will not call it
science if force is used to change belief, nor unless we can
discern some connection with our ability to predict and
control. But neither of these two criteria for the use of the
term 'science' suggest that the demarcation of science
from the rest of culture poses distinctively philosophical
problems.[33]

NOTES

1. Sometimes this meant simply that the rest of culture should
exemplify the moral virtues characteristic of the empirical scientist—
openness, curiosity, flexibility, an experimental attitude toward everything. Sometimes it meant, alas, that the rest of culture should adopt
something called "the scientific method." The former suggestion was
bracing and useful, but the latter led to ludicrous and unprofitable
self-criticism sessions, particularly among social scientists. I discuss
the relation between these two aspects of science-worship in "Pragmatism Without Method," in *Sidney Hook: Philosophy of Democracy and
Humanism,* ed. Paul Kurtz (Buffalo, N.Y.: Prometheus Books, 1983),
259–73.

2. See A. Fine, "The Natural Ontological Attitude," in *Scientific
Realism,* ed. J. Leplin (Berkeley: University of California Press, 1984),
83–107.

3. See my "Beyond Realism and Anti-Realism," in *Wo Steht die
Sprachanalytische Philosophie Heute?,* ed. Herta Nagl-Dockerl, et al. (Vi-
enna, 1986).

4. See Donald Davidson, "A Coherence Theory of Truth and
Knowledge," in *Kant oder Hegel?* ed. Dieter Henrich (Stuttgart: Klett-
Cotta, 1983), pp. 423–38; see p. 425.

5. See my "Solidarity or Objectivity?" in *Post-Analytic Philosophy,*
ed. J. Rajchman and C. West (New York: Columbia University Press,
1985), pp. 3–19.

6. For more on this point, see my "Pragmatism, Davidson and
Truth," in *Actions and Events: Perspectives on the Philosophy of Donald Davidson,* ed. E. LaPore and B. McLaughlin (New York: Blackwell,
1986), pp. 333–55.

7. D. Davidson, *Inquiries into Truth and Interpretation* (Oxford:
Clarendon Press, 1984), p. 194.

8. The latter claim is preserved in, for example, Richard Boyd's
"The Current Status of Scientific Realism," in *Scientific Realism,* ed. J.
Leplin, pp. 41–82; see p. 62.

9. Ernan McMullin, "A Case for Scientific Realism," in *Scientific Realism*, pp. 8–40; pp. 24–25.

10. Ibid., p. 14.

11. Ibid., p. 8.

12. Clark Glymour, "Explanation and Realism," in *Scientific Realism*, pp. 173–92; p. 173.

13. Boyd, "The Current Status of Scientific Realism," p. 66.

14. Ibid., p. 70. Boyd continues, "This is a nontrivial intellectual problem, as one may see by examining the various different attempts—behaviorist, reductionist and functionalist—to explain what a scientific foundation for psychology would look like." I quite agree that it is nontrivial, but I do not understand Boyd's example. This is because I do not see the connection of the debates between, e.g., Skinner and Chomsky, or Fodor and his opponents, to issues about scientificity.

15. Michael Levin, "What Kind of Explanation is Truth?" in *Scientific Realism*, pp. 124–39; p. 126.

16. Bernard Williams, *Ethics and the Limits of Philosophy* (Cambridge, Mass.: Harvard University Press, 1985), p. 136.

17. Ibid., p. 138.

18. Ibid., p. 139.

19. Ibid.

20. Ibid., p. 136.

21. Gilbert Harman, *The Nature of Morality* (New York: Oxford University Press, 1977), p. 6. See Thomas Nagel's discussion of this passage in *The View from Nowhere* (Oxford: Oxford University Press, 1986), p. 145.

22. See Mary Hesse, *Revolutions and Reconstruction in the Philosophy of Science* (Bloomington: Indiana University Press, 1980), pp. x–xi.

23. Ibid., p. 111. See also my "Unfamiliar Noises: Hesse and Davidson on Metaphor," *Proceedings of the Aristotelian Society*, Suppl. Vol., 61 (1987), 283–96.

24. Williams, *Ethics and the Limits of Philosophy*, p. 135.

25. Williams, "The Truth in Relativism," *Proceedings of the Aristotelian Society* 75 (1974–75), 215–28; reprinted in Williams' *Moral Luck* (Cambridge: Cambridge University Press, 1981), and also in, e.g., *Relativism: Cognitive and Moral*, ed. Michael Krausz and Jack Meiland (Notre Dame, Ind.: University of Notre Dame Press, 1982).

26. That is: "*A* cannot correctly say that *B* speaks truly in uttering *S* unless *A* could also say something tantamount to *S*." The question of the applicability of this principle is the question of whether all sentences are on a par in respect to truth or whether in some cases we can make use of locutions like "true for Teutonic knights, but not for me," "true for phlogiston theorists but not for me," etc. I would reject all

locutions of the latter sort, and fill their place with "consistent with
the beliefs and desires of . . . , but not with mine."

27. It is worth noting, in this connection, that Peirce regarded
pragmatism as a generalization of Berkeley's way of breaking down
the primary-secondary quality distinction. See his review of A. C. Fra-
ser's edition of Berkeley's works, reprinted in his *Collected Papers,* ed.
Arthur W. Burks (Cambridge, Mass.: Harvard University Press,
1966), vol. 7, 9–38.

28. Nagel, *The View from Nowhere,* p. 106.

29. See Davidson's claim that "there is no such thing as a language, not if a language is anything like what philosophers, at least,
have supposed," in his "A Nice Derangement of Epitaphs," in *Philosophical Grounds of Rationality,* ed. E. Richard Grandi (Oxford: Oxford
University Press, 1986), pp. 157–74. Compare Henry Staten, *Wittgenstein and Derrida* (Lincoln: University of Nebraska Press, 1984),
p. 20: "The deconstructive critique of language could even be
phrased as a *denial that there is a language.*"

30. Williams, *Ethics and the Limits of Philosophy,* p. 199.

31. Examples of such narratives are Hegel's *Phenomenology of Spirit,*
Nietzsche's *Twilight of the Idols,* Dewey's *The Quest for Certainty,* Heidegger's *The Question Concerning Technology,* and Blumenberg's *The Legiti-
macy of the Modern Age.*

32. I try to develop this connection between romanticism and
pragmatism in "The Contingency of Language" and "The Contingency of Selfhood," in *The London Review of Books,* vol. 8, nos. 7 and 8
(April–May, 1986). An opposing view, objecting to my attempt to
"becloud the sober insights of pragmatism with the Nietzschean pa-
thos of a *Lebensphilosophie* turned linguistic" is offered by Habermas.
(For this passage, see his *Der philosophische Diskurs der Moderne* (Frankfurt: Suhrkamp, 1985), p. 242. Habermas believes that "a partiality
for reason has a different status than any other commitment." See
Habermas: Autonomy and Solidarity: Interviews, ed. Peter Dews (London:
Verso, 1986), p. 51. I would like to substitute "a partiality for freedom," and in particular for freedom of thought and communication,
for "a partiality for reason." The difference may seem merely verbal,
but I think that it is more than that. It is the difference between
saying "let us defend liberal democracy by politically neutral accounts
of the nature of reason and science" and saying "let our philosophical
accounts of reason and science be corollaries of our commitments to
the customs and institutions of liberal democracy." The latter, "ethnocentric," approach seems to me more promising, since my holist view
of inquiry suggests that there are no politically neutral instruments to
use for defending political positions.

33. I am grateful to Paul Humphreys for helpful comments on an
earlier version of this paper.

SCIENTIFIC RATIONALITY
AND THE "STRONG PROGRAM"
IN THE SOCIOLOGY OF KNOWLEDGE

Thomas McCarthy

From 1930 to 1935 Karl Mannheim taught sociology at the University of Frankfurt, in close proximity to the Institute for Social Research. The members of the Institute were all the more concerned to make clear their intellectual distance from Mannheim's sociology of knowledge. Beginning with the reviews of *Ideology and Utopia* by Herbert Marcuse and Max Horkheimer in 1929 and 1930, they subjected this "new concept of ideology" to repeated criticism.[1] And with good reason: their critical theory shared with Mannheim's sociology of knowledge a recognition of the interdependence of society and culture, of the essential historicity of forms of life and thought, of the socially conditioned character of human knowledge. As critical theorists, however, they wished to hold on to a strong distinction between true and false consciousness and thus to avoid drawing relativistic conclusions from this desublimation of the spirit. Even though traditional philosophy's radical dichotomy between the ideal and the real could no longer be maintained, it was not necessary, they argued, to retreat to relativism; nor was it possible to avoid it, as Mannheim hoped to do, through a dynamic synthesis of socially conditioned perspectives. What was needed was a new concept of truth which while abandoning the traditional God's-Eye point of view retained the dichotomy between the valid and the invalid, albeit in a more modest, suitably human form.

In recent Anglo-American philosophy impulses to relativism have come once again from the sociology of knowledge (as well as, of course, from cultural anthropology and the history of ideas). A number of internal philosophical developments have also pulled in that direction: the growing influence of Wittgensteinian perspectives, the ongoing reception of Heideggerian thought, the spread of post-empiricist tendencies in the philosophy of science, the more recent but very rapid reception of poststructuralist ideas, among others. All of this has served to discredit subject-centered paradigms of reason, much as the similar considerations advanced by Hegel and Marx, Darwin and Freud, historicism and pragmatism, had done once before. And this time, in what Fred Dallmayr has called "the twilight of subjectivity," there have been numerous proclamations as well of the "end of philosophy"— sometimes only of Philosophy with a capital *P,* as Richard Rorty puts it—that is, of the search for ultimate foundations, *a priori* insights, a grasp of the totality—and sometimes of philosophy in any form, however chastened.[2]

Max Horkheimer and his colleagues at the Institute for Social Research responded to the earlier round of desublimation by calling not for an end but for a transformation (or *Aufhebung*) of philosophy: while it was indeed necessary to abandon the search for absolutes, it was, in their view, no less necessary to retain emphatic notions of reason and truth, autonomy and justice. These would have to be reconstituted in a deontologized, detranscendentalized universe of discourse, and this would require a cooperation among modes of inquiry previously kept separate. Thus was critical theory conceived as what Horkheimer called a "continuous dialectical interpenetration" of philosophy and empirical research, a "philosophically oriented historical and social research," in which philosophical concerns inspired empirical studies and the results of the latter worked back upon the conception of the former.[3] This approach had originally to be established in the wake of a temporalistic relativism fostered by historicism and in the face of a societal relationism advanced by the sociology of knowledge. The conditions under which it might today be renewed are not altogether dissimilar.

In this paper, I would like to consider one of the principal strains of contemporary relativism—the so-called strong program in the sociology of knowledge—and from a point of view broadly inspired by Horkheimer's earlier reaction to Mannheim. In line with the theme of these essays, I shall be focusing here on questions of scientific knowledge. If the strong program could succeed in that domain, its chances of succeeding in other domains seem very good indeed. In particular, I want to take up a question Ernan McMullin recently posed vis-à-vis the sociology of science: "If science is, at least in *some* sense, a social product . . . how can [its] rationality be said to transcend the society that produces it?"[4] And I want to defend the answer that the development of science has to be understood in some sense as a learning process. As will be evident, my concern here will not be with developments internal to the history of science but with the relation of scientific rationality, in a broad sense, to what I shall refer to, equally broadly, as magico-mythical thought.

1. Sociological Program cum Philosophical Rationale

Barry Barnes and David Bloor have recently advocated a "strong program" in the sociology of knowledge. 'Knowledge' here refers to "any collectively accepted system of beliefs" rather than to the "justified true belief"of the philosophers.[5] Eschewing any concern with the epistemic status of what is to be studied, they recommend treating all beliefs "on a par with one another as to the causes of their credibility."[6] On this approach, "the incidence of

all beliefs without exception calls for empirical investiga-
tion and must be accounted for by finding the specific
local causes of this credibility . . . without regard to the
status of the belief as it is judged and evaluated by the
sociologist's own standards."[7] We are simply to "investigate the contingent determinants of belief without regard
to whether the beliefs are true or the inferences rational."[8]
There is a powerful philosophical rationale behind this
abstinence:

> The relativist, like everyone else, is under the necessity to
> sort out beliefs, accepting some and rejecting others. He will
> naturally have preferences and these will typically coincide
> with those of others in his locality. The words 'true' and
> 'false' provide the idiom in which those evaluations are ex-
> pressed and the words 'rational' and 'irrational' will have a
> similar function. . . . The crucial point is that the relativist
> accepts that none of the justifications of his preferences can
> be formulated in absolute or context-independent terms. In
> the last analysis, he acknowledges that his justifications will
> stop at some principle or alleged matter of fact that only has
> local credibility. . . . For the relativist there is no sense at-
> tached to the idea that some standards or beliefs are really
> rational as distinct from merely locally accepted as such. Be-
> cause he thinks that there are no context-free or super-
> cultural norms of rationality, he does not see rationally and
> irrationally held beliefs as making up two distinct and quali-
> tatively different classes of things. They do not fall into two
> different natural kinds which make different sorts of appeal
> to the human mind, or stand in a different relationship to
> reality, or depend for their credibility on different patterns of
> social organization. Hence the relativist conclusion that they
> are to be explained in the same way.[9]

This undercuts the rationalist distinction between validity
and credibility and the accompanying claim that whereas
the latter might be explained by "contingent determinants," the former can be accounted for only by appeal to
reasons. There is nothing, Barnes and Bloor insist, so contingent and socially variable as what counts as a reason

for what. Something is evidence for something else only in a certain context, and thus this relation is itself a "prime target for sociological inquiry and explanation."[10]

Here the symbiosis between sociological program and philosophical viewpoint is evident. It is also typically involved, though seldom so explicitly, in the philosophical defense of relativism. There it is assumed that historians, sociologists, and anthropologists supply us with the descriptive accounts of the diversity of belief systems that furnish the empirical background to the philosophical argument. Any doubts about one partner in this intimate association will affect our confidence in the other: If philosophical relativism were to prove indefensible, we might well want to give the validity/credibility distinction some play in the sociology of knowledge; if evaluatively neutral, purely descriptive accounts of systems of belief proved to be impossible in principle, we might well come to doubt some of the usual arguments for philosophical relativism. In what follows I want to argue that something like this latter situation does in fact obtain, that the strong program in the sociology of knowledge is misconceived from the start.

2. Deabsolutizing the Notion of Truth

In his 1930 review of Mannheim's *Ideology and Utopia,* Horkheimer argued that the historically conditioned character of thought is not *per se* incompatible with truth. Only against the background of traditional ontological-theological conceptions of eternal, unchanging truth could this seem to be the case.[11] But there is no need, he went on, of any absolute guarantee in order to distinguish meaningfully between truth and error. Rather, what is required is a concept of truth consistent with our finitude, with our historicity, with the dependence of thought on changing social conditions. On such a concept, failure to measure up to absolute, unconditioned standards would be irrelevant. To regard this failure as leading directly to cognitive relativism is just another version of the "God is dead, everything is permitted" fallacy of disappointed expectations. As Horkheimer put it:

> That all our thoughts, true and false, depend on conditions which can change is just as certain as that the idea of an eternal truth outliving all knowing subjects is unfullfillable. This in no way affects the validity of science. . . . Anyone who is concerned about the correctness of his judgments about innerworldly objects has nothing to hope and nothing to fear from a decision on the problem of absolute truth. . . . It is not clear to me that the fact of *Seinsgebundenheit* (existential determination) should affect the truth of a judgment—why shouldn't insight be just as *seinsgebunden* as error?[12]

In short, there is no *direct* route from finitude and historicity to relativism, though there may well be a longer path. In the first instance, the challenge is to deontologize, detranscendentalize the notion of truth.

In his 1935 essay "On the Problem of Truth" Horkheimer attempted to do just that. He argued against the equation of fallibility with relativity. To grant that there is no final and conclusive theory of reality of which we are capable is not at all to abandon the distinction between truth and error. We make this distinction in relation to "the available means of cognition."[13] The claim that a belief is true must stand the test of experience and practice *in the present.* Knowing that we are fallible, that what stands the test today may well fail to do so in the next century, does not prevent us, or even exempt us, from making and defending claims to truth here and now. Consider the following: *A* claims that *S* is *p. B* disputes his claim. They resort to "experience and practice," the "available means of cognition," to settle the matter. This shows that *S* is indeed *p. A* was right. But *B* points out that there is no absolute warranty of truth. Experience and practice are historically variable. If the past is any guide, the means of cognition available in the next century will be different from those presently available. Therefore, *B* concludes, *A* should withdraw his claim to truth.

I think it is clear that this conclusion does not follow. *A*'s claim stands unless and until it is actually defeated. If *B*'s argument carried, it would entail the withdrawal of all truth claims, that is, the end of speech as we know it or even can imagine it. Or perhaps we should just attach the rider: "As far as we know" to every statement. But then it would not be doing any work at all. The practices of separating truth from error would continue undisturbed. What was reasonable to accept before adding the explicit reminder of our finitude and fallibility would be reasonable to accept thereafter. The abstract recognition that all our beliefs are open to correction does not make a rationally warranted belief any less warranted, any less rational. As Horkheimer put it: "A later correction does not mean that a former truth was formerly untrue."[14]

To get at this idea we might consider Hilary Putnam's treatment of 'truth' in terms of rational acceptability:

> Denying that it makes sense to ask whether our concepts 'match' something totally uncontaminated by conceptualization is one thing; but to hold that every conceptual system is therefore as good as any other would be something else. . . . To reject the idea that there is a coherent 'external' perspective, a theory that is simply true 'in itself', apart from all possible observers, is not to identify truth with rational acceptability. Truth cannot simply *be* rational acceptability for one fundamental reason; truth is supposed to be a property of a statement that cannot be lost, whereas justification can be lost. . . . Truth is an *idealization* of rational acceptability. We speak as if there were such things as epistemically ideal conditions, and we call a statement true if it would be justified under such conditions.[15]

There are various ways of capturing this idealizing moment in our conception of truth—e.g., Kant's notion of a regulative idea, Peirce's notion of the opinion fated to be ultimately agreed upon by all who investigate, or Habermas' notion of consensus arrived at in unconstrained rational discourse. For our present purpose, the differences between these formulations are unimportant. What is important is the common idea that any adequate account of truth as rational acceptability will have to capture not only its immanence—i.e., its socially situated character—but its transcendence as well. While we may have no idea of standards of rationality wholly independent of historically concrete languages and practices, it remains that reason serves as an ideal with reference to which we can criticize the standards we inherit. Though never divorced from social practices of justification, the idea of reason can never be reduced to any particular set of such practices. Correspondingly, the notion of truth, while essentially related to warranted assertibility by the standards or warrants of this or that culture, cannot be reduced to any particular set of standards or warrants. To put this another way, we can, and typically do, make historically situated and fallible claims to universal validity.

Applying this to Horkheimer's argument, we would have to alter his statement: "A later correction does not mean that a former truth was formerly untrue" to read: "A later correction does not mean that a formerly warranted belief was formerly unwarranted" or alternatively: "A later correction does not mean that a formerly justified claim to truth was formerly unjustified." That is, the belief, which we now know to be untrue, was warranted, justified, rationally acceptable in previous circumstances—but not ideally. We now have to take a closer look at this notion of "ideally." Can it mean anything more than "from our point of view," "according to our standards and criteria"? If we surrender, as Horkheimer does, the Hegelian notion that "we" can ever occupy the position of absolute knowledge, our judgment of "their" beliefs as rationally acceptable in those circumstances but not ideally can, it seems, *effectively* amount only to saying that what was acceptable to them is not to us. As Barnes and Bloor put this point, the distinction between validity and credibility is not an absolute but a "local" distinction relative to the accepted methods and assumptions of the evaluator's own group. To pretend otherwise is merely to suppose that credibility and validity are identical in one's own case.[16] So we seem to have gone round in a circle and come back to the very point at which we first encountered relativism.

3. Interpretation and Dialogue

Horkheimer's way of breaking out of this circle was to insist that the for us/for them relation is a dialectical one.[17] The invocation of "dialectic" here means, to begin with, that the relation of social investigators to the belief systems they investigate is not at all the relation of neutral observers to a world they describe from some Archimedean point outside of it; nor is it a relation of empathetic identification with subjects whose world can then be faithfully reexperienced. It is rather analogous to a dialogue relation between two subjects, "we" and "they," which allows not only for differences in point of view but also for argumentative adjudication of these differences. "For them" p is rationally acceptable: it is warranted by their experience and practice in light of their canons of reason and criteria of truth. "For us" it is not rationally acceptable: we detect "conditional and one-sided aspects" in their practices and standards. We attempt to "criticize" and "relativize" their views from our putatively more adequate point of view. It is important to keep in mind, however, that we may be mistaken as well as they. While we are convinced that p is not rationally acceptable and are prepared to make the case to that effect, we have to remain open to counterarguments from them, as well as from others.

Note that this model of critical dialogue corrects a certain imbalance in the images of the interpretive situation projected by the recently very influential "principle of charity in interpretation." In an understandable reaction to the ethnocentric biases of Victorian anthropology, contemporary interpretation theorists typically stress the obligation of the interpreter to broaden his or her horizon so as to make place for the very different concepts and beliefs, standards and practices of alien cultures and bygone epochs. Thus Winch, in his much discussed essay on "Understanding a Primitive Society," argues that "the onus is on us to extend our understanding" rather than "insist on seeing (things) in terms of our own ready made distinction(s)."[18] As a corrective to ethnocentrism, Winch's version of the charity principle is all to the good. But as an account of the logic of the interpretive situation it too is one-sided.

If there is initial disagreement in beliefs and practices, concepts and criteria, and if there is no extramundane standpoint from which we might neutrally adjudicate the differences, and if neither side is justified in assuming without further ado the superiority of their own way of looking at things, then a discussion of the differences is the only nonarbitrary path open for weighing the pros and cons of the divergent outlooks. If this discussion is to be symmetrical, then "their" views will have to be given equal consideration with "our" views. We shall have to try to appreciate how things look from *their* point of view. And this will require expanding the horizon of our own, as Winch points out. What he fails to consider is that symmetry requires the same from them. That is, *they* would, correspondingly, have to learn to see things from *our* point of view as well, if discussion of the differences in belief and practice is to proceed without unduly privileging the one or the other side. As I shall argue below, this description of the logical—or, in this case, dialogical—situation has far-reaching consequences for the whole discussion of relativism. (Note that this is meant to be an account of the *logic* implicit in the encounter between different worldviews and forms of life. In actual fact, of course, understanding alien cultures usually takes the form of a monologue, with imperialist overtones, in which "they" can gain a voice only to the extent that they learn to speak the hegemonic language, i.e., become "modernized.") The view I am recommending can perhaps be thought of as a kind of *Gedankenexperiment:* What would it be like if we had to defend the all-too-often merely taken-for-granted superiority of modern Western views in *symmetrical* dialogue with other cultures? What might we learn from them as well as they from us?

4. *Learning at the Socio-Cultural Level*

It might seem that this is merely to substitute a hermeneutic model of dialogue or conversation for the sociologist of knowledge's attitude of neutral description—and without having solved our original problem, for there are hermeneutic versions of relativism as well. On such accounts, interpretation is viewed as necessarily situation-bound, an understanding from a point of view, on the same level as what is understood. Thus social inquirers are not, as we may mistakenly suppose, neutral observers, explainers, predicters; neither are they sovereign critics who may safely assume their own cognitive superiority. They are, however virtually, partners in dialogue, participants in a conversation about the common concerns of human life.

Critical theorists need not, I think, deny this; but they would have to resist the relativistic implications that are usually drawn from it. To grant that every point of view is historically situated is not *ipso facto* to surrender all claim to validity, to drop any claim that one view is better than another. For instance, in the context relative to debating the relative merits of prescientific and scientific accounts of natural processes, the technological advances connected with the former would count as *one* argument in their favor. As embodied, active beings, our ordinary prescientific understandings of the world are, as Charles Taylor has recently put it, "inseparable from an ability to make one's way around in it and deal with the things in it," that is, from "recipes for action." Technological advances, as "more far-reaching recipes," cannot but force the issue between scientific and prescientific theories, for they "command attention and demand explanation" not only from our point of view but from theirs as well, that is from the point of view of any group that has to reproduce its existence in active exchange with the environment— that is, from the point of view of any human group.[19] Thus in our transcultural dialogue we might want to take the position that there has been a learning process in regard to our *technical* understanding of natural processes, that we have learned how to pursue the common human interest in prediction and control more effectively by differentiating its pursuit from other—moral, emotional, symbolic, aesthetic, etc.—concerns. In short, while the open-ended "conversation of mankind" rules out the assumption that our point of view is absolute, it does not require us simply to drop notions of cognitive advance or learning from experience.

On the other hand, it does not give us any general license to apply these notions wherever "their" views are different from "ours." We have to be prepared to learn from them in areas where their experience has taught them things we don't know, or know less well, or once knew but have since forgotten. Logically speaking, we have to allow for regression as well as progression in any given domain, as well as for straightforward differences in approach and differentials of development. While they may have something to learn from us regarding science and technology, we may have something to learn from them about alternative attitudes toward and conceptions of nature. While they may have something to learn from us about individualistic conceptions of self and personality, we may have something to learn from them about the limits to such conceptions.

There are, however, some important respects in which the learning situation may not be symmetrical. Consider the case of magico-mythical worldviews in tribal societies. If it is true, as anthropologists claim, that they are typically characterized by the lack of a developed awareness of alternatives to the established body of beliefs, and that this accounts in part for such phenomena as the sacredness of beliefs, taboo avoidance reactions in the face of challenges to them, and the magical power of words (which are regarded as having a unique and intimate relation to things)—then the very process of their coming to understand how we see things (apart from any discussion of relative merits) could have far-reaching effects on their views. For one thing they would have to understand in understanding us is our historically, sociologically, and anthropologically schooled view of the diversity of systems of belief and practice. To put what could be a very long story in a nutshell: the symmetry of the dialogue situation would require that they try to understand our beliefs and practices—as we must theirs—including the reasons why we hold the beliefs we do, the justifications we offer for accepting the practices we do, as well as the criticisms we have developed in rejecting other alternatives in our past—some of them rather similar to those obtaining in their society. I think it is clear that this very attempt to find a common language for discussing relative merits might well lead, independently of any specific discussions, at the very least, to a more qualified acceptance of those aspects of the tribal culture that depend precisely on *not* having had certain experiences—e.g., of cultural change and cultural pluralism—we have had, and on *not* having learned certain things we have learned—e.g., about the

historical and social variety of systems of belief and practice.

The same sort of consideration applies to other formal
features of what they would have to come to understand in
coming to understand our point of view—for instance, to
the differentiations we make between the concerns of science and religion, art and morality, and so forth. It is
typical of tribal and traditional systems of belief that they
do not differentiate in the same way, that they address
simultaneously what we would describe as religious, scientific, moral, emotional, and aesthetic concerns. They are,
in Alasdair MacIntyre's phrase, "poised in ambiguity,"
such that the very posing of such questions as: Is this
science or theology, symbolic expression or applied tech-
nique, or all at once? can permanently upset the
balance—as it did in our own past.[20] And again, to understand much of what we believe, they would have to understand the type of reflective, second- and higher-order
activities in which we engage—explicating and evaluating
basic categories and assumptions, aims, and criteria, considering alternatives and justifying choices—coming to
understand which might itself bring with it irreversible
changes in outlook. We too often forget, I think, that in
some situations it is difficult not to learn, and difficult
often to forget or ignore what one has learned.

Furthermore, once the discussion got underway, they
would have to face other deep-seated disadvantages. As
MacIntyre, following Hegel, has recently argued, in making the case for the rational superiority of a view to its
rivals, one very important consideration is its ability to
explain the accomplishments and failures of its rivals, to
incorporate their strengths and transcend their limitations.[21] While this tactic may lead in the case of contemporary debates to a situation in which, as Richard Rorty
has put it, everyone goes around applying an *Aufhebung* to
everyone else, its consequences in the case at hand are far

more one-sided. For offering an alternative account of our science and technology, law and morality, art and religion, etc., incorporating their accomplishments and transcending their limitations is evidently going to be more of a problem for them than the corresponding task is for us—for our culture has been involved in doing just this sort of thing for some time now. For them, however, this would be, in a way, to cease being who they are. This is not to say that we are always right and they are always wrong. But it is to suggest that the concept of learning has its application at the socio-cultural level as well, and that as a result the conversation of humankind does not place each system of beliefs on a par with every other. Some differences are more than mere differences, precisely because they can best be understood as the results of learning.

5. Neutrality in the Sociology of Knowledge

Here the advocates of a strong program in the sociology of knowledge might take a different tack; they might grant this point and argue that while nothing *prevents* an interpreter from adopting the attitude of a participant in dialogue or argumentation, nothing *requires* him or her to do so. They might thus propose a division of labor between the sociologist of knowledge who describes and explains belief systems and the humanist or critical theorist who evaluates and criticizes them—that is, between investigators who adopt the objectivizing attitude of the non-involved and those who adopt the performative attitude of participants. Max Horkheimer and his colleagues repeatedly argued that this alternative was an illusion, that the social investigator was unavoidably "involved" in the object he or she was studying, that the sociologist of knowledge's attitude of above-the-battle neutrality was a prime instance of false consciousness. Arguments to this

effect have turned up in the contemporary relativism debates as well. It has been argued, for instance, that in
explaining actors' beliefs, sociologists cannot avoid en-
dorsing or rejecting the reasons the actors would give for
them.[22] If they are pursuing a strong program in the manner of Barnes and Bloor, then they offer explanations in
terms of such "contingent determinants" as processes of
socialization and class membership, social integration and
cultural transmission, influences of authority, roles, institutions, group interests, and so forth. The variety of candidates for "contingent determinants of belief" is
obviously very wide. There does not appear to be any
property they possess in common, except perhaps that
they are *not* the sorts of evidencing reasons actors *themselves*
would give for their beliefs.

It is crucial at this point in the argument to note that
Barnes and Bloor want to avoid opposing social determinants, on the one side, to epistemic factors or evidencing
reasons, on the other. (It is difficult to imagine how the
"incidence of belief" in this or that scientific theory could
be explained without reference to reasons *in some form.*)
Instead, they argue that evidencing reasons *are a species of*
social determinant. "It would be difficult," they write, "to
find a commodity more contingent and more socially variable than . . . evidencing reasons."[23] In short, in the relation, '*x* is a reason for *y*', what counts as a good reason for
holding a belief or performing an action is socially variable and socially explicable. Their point, then, is not that
reasons cannot figure in the accounts given by the sociologist of knowledge but that they are referred to descrip-
tively and not normatively, that they are mentioned but
not used.[24] No beliefs are explained in a way that presupposes a normative evaluation of them. In particular, the
sociologists' own evaluations play no role in their explanations. Explanation is perfectly symmetrical in respect to
true and false beliefs, rational and irrational practices—as
they would judge these things from their own socially situated points of view.

It is this claim, the putative neutrality of the sociologist of knowledge, that I want to consider in the remainder of this paper. I shall suggest, as Horkheimer did fifty years ago, that this is a subtle instance of self-misunderstanding, that the evaluations given by the sociologist of knowledge do play a role, however implicitly, in their explanations. The argument, which I can only sketch here, runs as follows. Everyone seems agreed, finally, that we have to understand beliefs and practices in their settings and uses. The very identification of what is to be explained is an interpretive enterprise, and this inevitably lands us in a hermeneutic circle. For my purposes, it is not important whether this circle is explicated in terms of Gadamer's *Vorgriff auf Vollkommenheit,* Winch's principle of charity in interpretation, Davidson's version thereof (according to which "we must assume that by and large a speaker we do not yet understand is consistent and correct in his beliefs—according to our own standards, of course,"[25]) or Richard Grandy's weaker "principle of humanity" to the effect that whether we regard their beliefs as true or false, *their being held* must be intelligible.[26] What interests me here is the common point that meaning and validity are inseparable, that the identification and interpretation of beliefs is possible only under certain pragmatic constraints to the effect that we view what we are interpreting as largely reasonable in its own context. Most relativists are only too happy to grant this notion of "reasonableness in context," for it appears to be just the sort of local notion they favor. But they fail to see that there is a dialectical pitfall here, which Alasdair MacIntyre pointed out some time ago in his critique of Winch.[27] For, *to whom* must the behavior be shown to be intelligible or reasonable in its context? To the interpreters, of course (not to the natives, surely). And this means: by the light of standards of intelligibility and rationality amenable to them. The interpreters must come to see how a certain belief or practice could appear reasonable in a certain context. If the belief is one their culture does not hold, they will look for differences of context to account for the differences of belief. That is, they will note features present in their context and absent from ours, or present in our context and absent from theirs, such that these presences and absences make the differences intelligible. Notice that all of this is required *in the very attempt to understand* alien systems of belief. Notice also that ascertaining the *what* of belief has also and inescapably involved us in grasping the *how* and *why* of belief. Notice finally that the interpreters' involvement in this process is such that their evaluations do play a role, and an essential role, in their account of alien beliefs. It is their standards that are at work in their sense of what might reasonably be believed in a given context, of what might be accounted for by what they regard as normal processes of perception, cognition, and reasoning, and of what can be accounted for only by pointing to special circumstances.

The dialectical point is that our explication is already, at least implicitly, an evaluation. It says, in effect, that a given belief might make sense and be reasonably held in a given context—for example, in the context of a preliterate, relatively stable and homogeneous, relatively isolated culture, in the absence of a developed science, philosophy, historiography, etc.—and this amounts to saying that it is a belief which could no longer reasonably be held. Or we may find that what we have to abstract from in our setting, or notice in their setting, in order to make a difference in belief understandable, makes us aware for the first time of the peculiarities of certain aspects of our context, and thus of the deficiencies in the beliefs that we now see to be limited by them. My point here is not to decide who is right and who is wrong on any particular matters of belief, but rather to demonstrate that the position of the sociologists of knowledge is not that of the detached spectator but that of the participating player. They are implicitly involved in evaluating beliefs even when they are trying merely to describe and explain them. For meaning and validity are intertwined and both are internally connected with reasons. In interpreting, we cannot but move in the space of reasons. And in coming to understand "their" beliefs and how "they" reason about them, the question of whether or not they have come to hold them for reasons "we" can accept cannot but play a role in how we proceed. Furthermore, when we offer an account of their beliefs which differs from their own account we have *eo ipso* criticized them, implied that we are right and they are wrong.

Critical theorists make this implicit dialectic explicit. They openly address the question of the validity of the belief systems they study. And thus they avoid the self-misunderstanding that still—fifty years later—plagues the sociology of knowledge and those philosophers who draw uncritically upon it.

NOTES

1. H. Marcuse, "Zur Wahrheitsproblematik der soziologischen Methode," *Die Gesellschaft* 6 (1929); M. Horkheimer, "Ein neuer Ideologiebegriff?" *Archiv für die Geschichte des Sozialismus und der Arbeiterbewegung* 15 (1930), reprinted in V. Meja, N. Stehr, eds., *Der Streit um die Wissenssoziologie*, 2 vols. (Frankfurt: Suhrkamp, 1982), 2:474–96.

2. See K. Baynes, J. Bohman, T. McCarthy, eds., *After Philosophy* (Cambridge, Mass.: MIT Press, 1986) for an overview of the issues; Rorty's essay, "Pragmatism and Philosophy," appears there on pp. 26–46.

3. M. Horkheimer, "Die gegenwärtige Lage der Sozialphilosophie und die Aufgabe eines Instituts für Sozialforschung," *Frankfurter Universitätsreden*, Heft XXXVII (1931); reprinted in M. Horkheimer, *Sozialphilosophische Studien* (Frankfurt: Fischer, 1972), pp. 33–46; here pp. 40, 45.

4. E. McMullin, "The Rational and the Social in the History of Science," in J. R. Brown, ed., *Scientific Rationality: The Sociological Turn* (Dordrecht: Reidel, 1984), pp. 127–63; here p. 127.

5. B. Barnes, D. Bloor, "Relativism, Rationalism and the Sociology of Knowledge," in M. Hollis, S. Lukes, eds., *Rationality and Relativism* (Cambridge, Mass.: MIT Press, 1982), pp. 21–47; here p. 22.

6. Ibid., p. 23.

7. Ibid.

8. Ibid.

9. Ibid., pp. 27–28.

10. Ibid., p. 29.

11. "Ein neuer Ideologiebegriff?" p. 485.

12. Ibid., pp. 485–86.

13. M. Horkheimer, "On the Problem of Truth," in A. Arato, E. Gebhardt, eds., *The Essential Frankfurt School Reader* (New York: Vrizen Books, 1978), pp. 407–43; here p. 421.

14. Ibid., p. 422.

15. H. Putnam, *Reason, Truth and History* (Cambridge: Cambridge University Press, 1981), pp. 54–55.

16. "Relativism, Rationalism, and the Sociology of Knowledge," p. 30.

17. I do not claim to be faithful to Horkheimer's understanding of 'dialectic' in what follows.

18. P. Winch, "Understanding a Primitive Society," in Bryan Wilson, ed., *Rationality* (New York: Basil Blackwell, 1971), pp. 78–111; here p. 102.

19. Charles Taylor, "Rationality," in Hollis and Lukes, eds., *Rationality and Relativism*, pp. 87–105; here pp. 101–3. Compare Habermas' formulation of this argument in terms of basic, species-wide cognitive interests, in *Knowledge and Human Interests* (Boston: Beacon, 1971).

20. A. MacIntyre, "Rationality and the Explanation of Action," in A. MacIntyre, *Against the Self-Images of the Age* (London, 1971; rpt. Notre Dame, Ind.: University of Notre Dame Press, 1978), pp. 244–259; here p. 252.

21. A. MacIntyre, *After Virtue*, 2nd ed. (Notre Dame, Ind.: University of Notre Dame Press, 1984), Postscript, pp. 265–72.

22. See J. Habermas, *The Theory of Communicative Action*, vol. 1: *Reason and the Rationalization of Society* (Boston: Beacon, 1984), pp. 102–36; my critique of his argument, "Reflection on Rationalization in the *Theory of Communicative Action*," in R. Bernstein, ed., *Habermas and Modernity* (Cambridge, Mass.: MIT Press, 1985), pp. 177–91; here pp. 183–86; and his response, "Questions and Counterquestions," in Bernstein, ed., *Habermas and Modernity,* pp. 192–216, here pp. 203–5. The argument of the present paper represents a shift in the position I advanced in my critique.

23. Barnes and Bloor, "Relativism, Rationalism and the Sociology of Knowledge," p. 28.

24. See Gary Gutting, "The Strong Program: A Dialogue," in J. R. Brown, ed., *Scientific Rationality: The Sociological Turn*, pp. 95–112.

25. D. Davidson, "Psychology as Philosophy," in *Essays on Actions and Events* (Oxford: Oxford University Press, 1980), pp. 229–44; here p. 238.

26. R. Grandy, "Reference, Meaning and Belief," *Journal of Philosophy* 70 (1973): 439–52. Steven Lukes runs through some of these different versions in "Relativism in Its Place," in Hollis and Lukes, eds., *Rationality and Relativism*, pp. 261–305.

27. A. MacIntyre, "Is Understanding a Religion Compatible with Believing?" in B. Wilson, ed., *Rationality*, pp. 62–77.

SOCIALIZING EPISTEMOLOGY

Mary Hesse

1. Epistemology of Science

Quine coined the phrase "epistemology naturalized"[1] to express his rejection of the view, shared by classical rationalism and by Kant, that there are fixed *a priori* truths about logic or conceptual frameworks to which science must conform, and that it is by these as well as empirical data that we have scientific knowledge. For Quine epistemology itself becomes an empirical study—it describes how we get around in our natural environment by inductive methods, which themselves are understood as extensions of animal learning. By means of trial and error, aided by both biological and cultural evolution, we get increasingly good knowledge of the natural world, as tested by the accumulating success of our predictions and capacity to control. The *a priori* elements previously identified by rationalists become contingent universals of human biology and psychology whose success has been guaranteed by the various mechanisms of selection and survival. One could speak of the "objectivized" model of the world as a construct of human sensory interaction with the real world. The position is still realist, but the ontology is not in *correspondence* with knowledge, rather there is a unified view of interaction within a natural reality embracing the physical, the biological, the psychological, and the human knower.

97

Quinean epistemology is essentially *individualist*. In spite of the fact that much is made of the importance of language as the chief agent of categorization, that is, as the theoretical framework within which instrumental coping with the natural world is organized, language is not conceived as a social institution—it remains detached from the evolutionary story and unexplained by it. It is an accidental accompaniment of human interaction with the natural world. There is indeed a tension between the biological metaphors with which human knowledge gets assimilated to animal knowledge on the one hand, and linguistic metaphors such as *conceptual framework, language games,* and *theoretical meaning,* which are essentially social. Animals do not, in the required sense, have language, and hints toward a cultural as well as biological theory of evolution are as yet blank checks, certainly not sufficient to take account of the social context of science and its language.

This is one reason why naturalized epistemology is not backed by much empirical evidence. The sociobiology of human communities and their language is not sufficiently developed (and perhaps never can be) to permit a merely biological account of human knowledge which develops and extends theories of animal learning. Other difficulties with naturalized epistemology are that the complexities of scientific development and its technical exploitation do not look, on the face of it, like the natural extension by the human species of animal learning mechanisms. Nor is it obvious that they have much to do with the survival of the species in the biological sense, at present indeed rather the reverse. To take account of such objections epistemology needs not only to be naturalized but also in some sense socialized.

The socialization of epistemology may be said to begin with Durkheim, whose overt aim was to replace the rationalist element in received epistemology by social categories. This is apparent first in his work with Mauss,

Primitive Classification, where the aim of socializing rational norms is explicit. Even in the later *Elementary Forms of the Religious Life,* which is best known for Durkheim's socializing of the categories of religion, the program of socializing epistemology occurs in the first chapter and is clearly intended as the presupposed framework for a particular application to religion.[2]

Durkheim notices the vast variety of categorial frameworks and beliefs operative in traditional societies and wishes to explain these, not by Levy Bruhl's "pre-logical mentality," but as symbolic elaborations of social structures. These elaborations take off from the social base and become quasi-autonomous systems of reasoning, appearing to claim the absolute allegiance of the mind. How is it, he asks, that logic, and the concepts of truth, cause, class, number, space, and time in a given society, attain consensus, stability, authority, and a quasi-absolute status? It can only be, he replies, that like the concept of God in religious societies, they have behind them social power and sanction, and these are derived from the actual force inhering in social norms and structures—clan divisions and hierarchies, seasonal imperatives of hunting and agriculture, and the like. Durkheim, however, like Levy Bruhl, declined to apply these insights to the modern scientific mentality, and so although he has a sociology of the rational and religion, he has no sociology of science. There are two chief reasons for this. The first is that his whole theory is predicated on an evolutionary view of human history, which itself rests on a scientific theory of natural evolution, and this theory cannot itself be socialized without apparently undermining its own credentials. In other words, as far as science is concerned, Durkheim remains a naturalized epistemologist. The second reason is that he also wishes to naturalize sociology: his sociological method is always explicitly positivist, and he believes that positivist method is the highest product of the evolving forms of knowledge, and is therefore, at least at our stage of evolution, effectively uncriticizable.[3] Positivist method, as well as scientific evolution theory, is presupposed for the sociology of knowledge *other than* the scientific.

In the end Durkheim's socialized epistemology is conservative and uncritical with respect to both natural and social science. He provides tools for the study of systems claimed as cognitive by other, nonscientific cultures, but neglects the self-reflexive critique of his own tools, and consequently the possibility of alternative modes of knowledge open to our culture. Meanwhile, precisely these alternative possibilities were being explored by hermeneutic studies of historical knowledge. Dilthey asks the epistemological question about the study of history that parallels the one Kant had asked about the study of nature: we *have* historical knowledge (in the flourishing studies of nineteenth-century Germany); *how* is it that we have it? Dilthey, Weber, Gadamer, Habermas, and others have answered in terms of various forms of *Verstehen,* and in terms of social *interests* in a form of knowledge that is not that of natural science but has to do with the practice of human communication and mutual interpretation. In other words, the normative knowledge of history, politics, and morals becomes the subject matter of a nonpositivist epistemology.

Modern Durkheimians have objected to Durkheim's exclusion of natural science from his socialized epistemology, but they have not always broken free of his positivism. For example, in their different ways both Mary Douglas and David Bloor[4] have tended to regard natural scientific methods (the search for testable laws, deductive explanations, etc.) as the appropriate methods for the sociology of natural science as well as for its own practice. But the symmetry thesis of Bloor's "strong program" asks that science should be subject to the same scrutiny as every other social institution. This surely means that the sociological method adopted in studying science should also be subject to self-reflexive scrutiny, in two respects.

First, the development of sociology of science itself needs a social explanation. This has been held to lead to vicious circularity or regress of explanation. I have argued elsewhere[5] that this need not be so from the standpoint of the strong program itself, and I shall not pursue that issue here. More important for our purposes is that the methods of sociology should not be presupposed to be uniquely those of the natural sciences. Recently adherents of the strong program have explicitly acknowledged normal *historical* criteria of adequacy as appropriate for sociology of scientific knowledge, and they look not only for generalizable causal laws, but also for the particular cause-effect relations and appeals to human intentions that characterize general history.[6] Thus sociology of science becomes to some extent itself a hermeneutic science.

In his strong program Bloor proposed that explanations in sociology of science should be symmetrical as between true and false theories, and between modes of thought which may or may not reflect twentieth-century Western interpretations of scientific method. Another way of putting this thesis is to say that scientific theory, past or present, should be investigated on the same basis as that adopted for the symbolic belief systems of cultures in general, according to the methods of sociology and social anthropology. This means, among other things, that cognitive terminology such as 'belief', 'knowledge', 'truth', 'falsity', 'rationality', will be used in describing actors' thoughts and behavior as it is used by actors themselves, without bringing any external judgments to bear from our own standards of knowledge and truth. It also means that social causation of beliefs will be sought in the same way whether we judge the beliefs to be true or false, rational or irrational, and that when this social methodology is applied to the past or present science of our own culture, even there no such judgments are relevant.

An unnecessary air of paradox may attach itself to this formulation of the strong program. If we take the sociolog-

ical analogy seriously, it is clear that the principle of sym-
metric explanation can be used to safeguard, rather than
to reject, the distinctiveness of Western science. For if we
consider, say, an anthropologist fallen among Cavendish
physicists, a necessary part of his method will be to learn
the institutional rules and norms of his subject matter,
and particularly whether there are distinctive standards
and rules of truth, correctness, empirical test, rational argument, and the rest, exhibited by his actors, and he will
have to try to understand these as they do. There is no
need, therefore, for the strong program to presuppose that
there is no difference between tentative scientific beliefs,
serious hypotheses, theories, and accepted parts of scientific knowledge, because these things may be governed by
the actors' own rules, and the rules will be different from
the overt criteria of truth and inference in traditional cultures. The institutional rules may or may not be identical
with those discussed by philosophers of science, but the
adherent of the strong program need not deny that in our
culture there is some recognition of technical and logical
"reasons" for adoption of the theories we do adopt. What
he may rightly go on to point out is that the actors' accounts may contain elements of false consciousness, and
that in any case adoption of the criteria themselves is
subject to social causation which can be investigated socially and historically.

I shall return to the question of science as an institution
with rules and norms, but meanwhile to fix ideas let us
look at two examples of social historiography of science. I
take these from physics during the last hundred years,
because it is in physics rather than in softer sciences that
the thesis of social constraint will be most difficult to establish. The examples are Paul Forman's now classic discussion of "Weimar Culture, Causality and Quantum
Theory," and Brian Wynne's interpretation of ether theories in Cambridge in the late nineteenth century in "Physics and Psychics: Science, Symbolic Action and Social
Order in Late Victorian England."[7]

Forman describes the introduction of acausality into quantum physics by German physicists in the aftermath of defeat in the First World War. His thesis is that this has to be explained in terms of contemporary attacks upon mechanistic science, coming from a romantically influenced intellectual milieu, and disseminated widely among physicists. Moreover, Forman argues that the rejection of causality in physics is not explicable in terms of internal scientific developments. In his paper, Wynne describes the intellectual milieu in which ether theories arose, and argues that this was one of conflict between, on the one hand, utilitarian and professional exploitation of materialist science as a basis for education and morals, and on the other hand a more conservative scientific realism for which such exploitation was seen as a threat to social order. In the conservative view which dominated Cambridge science, metaphysical and spiritual interpretations of human nature were seen as being supported by the postulate of a nonmaterial ether, and indeed ultimately by an ether which subsumed the theory of mechanical matter itself and thus provided a physical model of unity and continuity. Also, some of the Cambridge physicists were active in the nascent Society for Psychical Research, for which a scientifically grounded ether theory provided welcome intellectual support, and which was used to give rational backing to a less articulate spirituality of the uneducated classes. Like Forman, Wynne claims that the transformation of physical theory involved in the ether postulate was not required by the data nor by the tradition of physics. Moreover the postulate is not found among other physicists. Those working, for example, in contemporary Germany adopted the contrary postulate of action at a distance. Therefore, Wynne concludes, the appearance of the ether in a particular place and time must be related causally to nonscientific factors.

Both case studies make the historical claim that a theoretical discontinuity having important metaphysical implications cannot be explained as an outcome of purely scientific reasoning. "Purely scientific reasoning" might of course be interpreted in such a strict sense as to make this claim merely platitudinous, for it is generally accepted that in a complex scientific situation theories will be underdetermined by logic and evidence and hence not explicable by purely scientific reasons in that sense. But the claim becomes substantial if we take it in the sense that internal reasons drawn from logic, evidence, normal inductive reasoning, and the local *scientific* tradition ("background knowledge") are not sufficient, and moreover that the remaining explanatory gap cannot be filled by reference to individual psychology ("great man" theories of explanation), and should not be filled by appeal to simple historical accident.

Given that understanding, both authors present their case studies as standard history, open to historical debate and testing and perhaps refutation. Forman has received such a response from John Hendry,[8] who argues that there were after all sufficient internal reasons for change, and that in any case the rejection of causality was not in itself seen by physicists as the major conceptual shift of the new theory: it was less important, for example, than the reinterpreted relation between observer and observed, and the new space-time concepts implied by quantum theory.

Hendry concludes his paper with a series of questions about the nature of historical causation, and recognizes that in a situation as complex as Weimar science there is unlikely to be a simple set of causal influences, nor are these likely to be entirely explicit, either in the minds of the protagonists or in the historical sources available. In pursuing various examples of the lively debate at present going on in history of science between social historians and their critics, it becomes increasingly clear that first-order historical work is essential to argument about the validity of socialized epistemology, yet at the same time issues of principle are not going to be conclusively settled by historical evidence. The issues soon become deeper ones about historical methodology and about the nature of scientific knowledge. Protagonists on the two sides of the debate no longer dispute the *possibility* of a variety of causal factors from economic, political, social, intellectual, rational, and empirical sources; the dispute is rather between those who wish to *weight* these factors differently. Even rationalists can agree to a measure of socializing of their epistemology. They regard the history of science as primarily the arena of rational thought and experiment and wish to pursue rational explanation as far as it will go. But they can admit that such explanations may fail, as in archaic science, deviant science, or simply because of the underdetermination of theories by data, and then they may admit ideological, social, and psychological factors as causative. At the other extreme there are constructivists, who hold that social causation is overriding, the "rationality" of different cultures is an epiphenomenon of social factors, and that even the empirical world has little to do with the structures of scientific knowledge that emerge. To quote Harry Collins, "the natural world has a *small or nonexistent* role in the construction of scientific knowledge."[9] In the middle lies the strong program which need not deny natural as well as social causation although it holds that these explanations refer to "true" and "false" science symmetrically.

Thus a variety of epistemologies of science informs its history and sociology. But it looks as if we have now gone full circle from arguing that the epistemology of science is a subject for social and historical explanation, to the thesis that every history or sociology *presupposes* an epistemology. We should not be surprised at this circularity, which is no more vicious than the familiar circle of theory and observation: observations are theory-laden and theories are ultimately constrained by observation. In that case the circle is broken (at least according to most moderate philosophies of science) by retaining an element of empirical constraint, tested by the success of theories in organizing

and predicting experience in pragmatic ways that are
common to different theoretical interpretations. In the
present case the testing and choice of epistemologies of
science is more difficult and less decisive, but (again for
most moderate philosophies) the circle can be broken by
showing how a coherent epistemology fits the relevant intellectual, social, and historical factors, and judging which
epistemology fits them best.

2. Cognitive Systems

Sociology of science has often adopted models of
method from the social anthropology of general cognitive
systems. It is no accident that anthropology should be a
source of models, rather than the other human sciences.[10]
All historians and sociologists of ideas have conceptual
problems that require them to have an epistemology of
their subject matter, but students of Western thought have
generally been content to adopt models of rationality
based on Western logic and science as the paradigm
norms of thought. It is social anthropologists who have
most obviously been forced to consider modes of thought
different from Western rationality, and to take them seriously as cognitive systems. Since it is now the Western
tradition of scientific knowledge that is itself in question,
sociologists of science have sought alternative epistemologies particularly in the study of non-Western cognition.

For an epistemology of science based on the strong program the content of knowledge is no longer a datum that
awaits rational justification. The question arises therefore
as to what *is* the subject matter of socialized epistemology.
If all cognitive systems are to be treated symmetrically,
what are the marks of a specifically *cognitive* system as a
social institution? Social anthropology has adopted various approaches to this question.

There is first the *actors' model*, which attempts to do
without any cognitive judgments at all on the part of the

investigator, and is perhaps closest in spirit to the symmetry thesis of the strong program. The actors' model is the principle of giving privilege to actors' overt points of view. Forman's and Wynne's papers, and most standard social history of science, are exercises in this mode, in that they are concerned with the views and beliefs more or less explicitly presented by actors and in some cases argued for by them. A cognitive system is that which commands the explicit belief of most, or at least a significant number of members or subgroups of a society, together with institutional structures, performances, and actions that are associated with the set of beliefs. Belief tends to be construed as an individual state of mind, to be elicited by questioning, or by inference from actions, presupposing those actions to be (in our sense!) rationally based.

There are many well-known difficulties with the notion of actors' privilege. First, as philosophers of science have discovered in trying to elicit statements of method from scientists, it is not possible to follow the principle of privilege very far without simplifying, idealizing, and reconstructing. Actors' overt points of view are rarely systematic enough to provide a coherent story, and their accounts of belief do not always fit with observable actions. An epistemology is needed to discover an epistemology. Again, the notion of "belief" as an individual mental state has been criticized both in Western philosophy, and in regard to the interpretation of alien cultures. Rodney Needham's comparative study[11] showed that the Western concept of "belief" is not closely translateable into the concepts of other cultures, although there is no doubt that they do have cognitive systems in the sense of institutionalized religious mythologies and world models. As for the inference of beliefs from observable actions and goals, this method rests essentially upon Western conceptions of what is "rational action,"[12] and is therefore questionable both as a method of eliciting the cognitive systems of other cultures, and as a method for the sociology of science, where we

cannot initially presuppose what the rational goals of science are. In respect of science, this approach is in fact the
rationalist one, not the method of the strong program.

In the light of such objections, social anthropologists
have adopted *symbolic action* models for cognitive systems.
Cognitive systems are then the kinds of social institution
that are typically exhibited by a religion, including cultural orthodoxies, dogmas, and norms: "symbolic statements about the social order,"[13] as Leach puts it, which
issue in ritual utterances and actions. The social constructivist view of Western science is an attempt to construe
science similarly, as more like the myths and religions of
other societies than rationalists would countenance. In the
title of Wynne's paper there is a reference to "symbolic
action," although his actual discussion is more concerned
with actors' beliefs and face-to-face interactions. But he
describes a symbolic action model as one in which the
historian will recognize "tacit meanings . . . in a process
of continual renegotiation, defense and development,"
and in which the elaboration of symbolic universes will
depend more on what is perceived to be efficacious social
weaponry than on consistency with either social practice
or scientific logic.[14]

The "scientific worldview" of modern societies does
indeed have many of the features of a traditional symbolic
system. It is the taken-for-granted consensus, to be
wheeled out for purposes of pedagogy, ideology, and social
legitimation and persuasion, and to give factual grounding for action in relation to such things as medicine, ecol-
ogy, and nuclear power. But the analogy is otherwise not
close. There are not only the traditional criteria, such as
falsifiability and accumulation of pragmatic success,
which are held to demarcate science, but also social differ-
entiae, such as the open-endedness of scientific norms in
modern societies emphasized by Horton, or the atomistic
division of labor within science and within society to
which Gellner draws attention.[15] In particular, science has

not (except in maverick cases like Auguste Comte) issued in popular ritual celebrations which structure the daily life and actions of the majority of social individuals. It constitutes one great complex of social institutions among others, and makes no *official* claims to all-pervasive competence or authority. Rather the reverse, because we are used to hearing pleas from its practitioners for more study of the moral aspects of life in order to supplement the factual information in terms of which decisions have to be made. Official science is in fact generally *resistant* to the view that science is a "symbolic statement about reality" which in principle mirrors and directs every aspect of social and individual life.

For purposes of the social understanding of science it seems better to draw the boundaries of cognitive system more narrowly than an all-embracing symbolism. To do this we may refer to the cognitive sciences themselves. Studies of mental modeling of the world are clearly relevant to the ways in which both science and symbolism enable humans and animals to survive and attempt to realize their purposes. Such mental modeling is not mere copying. It is already evident from work on elementary perception that no one-to-one correspondence theory of truth provides an adequate theory of modeling even of the natural environment. "Mental modeling" must be taken to include structures having various analogical and morphic relations to "the world," both natural and social, which may require elaborate transformations and mediations through action before "the world as experienced" can be reconstructed. Moreover, models may be hypotheses, or "stories" about the world, which issue in testable action. The "success" of a model may be tested by comparing with further experiences the expectations induced by it, or by longer-term methods of biological selection, or by acceptability by social consensus and by the realization by societies of overall purposes both manifest and latent.[16]

So far, little is known about how mental schemas which

are social rather than individual (Durkheim's "collective representations" or Leach's "symbolic statements") work as cognitive systems. Functionalism, structuralism, and sociobiology are all theories of their operation, and none has been found generally adequate. But one might venture a definition of "cognitive system" along the following lines:

> A *cognitive system* is a collective mental schema modeling some aspects of the world, issuing in symbolic ritual and/or appropriate action to realize human purposes.

Such a definition has the merit of including those cognitive systems that are studied as such by both social anthropologists and cognitive scientists. It does not presuppose functionalism, structuralism, or sociobiology, because it does not assume that societies or the human species have latent purposes or common psychological structures or goal-directed fitness for survival. It does, however, place the cognitive within the domain of human intentionality. Does reference to "modeling the world" do enough to demarcate the cognitive from other human institutions? It seems, for example, to cover some aspects of the aesthetic that are not usually regarded as cognitive. Deciding what the limits of the cognitive are would require a lengthy discussion of whether "the world" that is modeled presupposes some form of scientific or other realism, or whether world-models are rightly regarded as cognitive if they answer in a more general sense to practical social needs. For present purposes it is enough to assume that the institutions of science are cognitive in the sense just defined. This is a weaker sense of 'cognitive' than is usually assumed for science, because it is consistent with the strong program and does not imply realism or exclusively rationalist justifications.

In the rest of this paper I shall try to bring out some of the issues involved in judging competing social epistemologies of science, and in particular I shall consider and reject an extreme version of constructivism. At the same time I hope to show that the epistemology of science may be socialized without neglecting the features that differentiate science from other cognitive systems.

3. Constructivism

The study of generalized cognitive systems incorporates empirical and nonempirical, natural and social knowledge alike. Constructivism is the particular social epistemology of scientific knowledge that attempts to assimilate it totally to the nonempirical and the social. In using the word 'knowledge' one has constantly to keep in mind that for the consuctivist it means simply 'cognitive system', and that when a constructivist speaks of science as an "institution for generating knowledge about the natural world," this apparently realist locution must not be taken at its face value.

Paraphrasing, the constructivist means something like the following: "Our society has developed an institution that we call science, whose product is 'knowledge about the natural world'. This product of science is the way our society agrees, because of social conventions and training, to see the natural world. It is the best way we have of arriving at consensus about it. Other peoples have other ways: Zande witchcraft, Lele pangolin cults, Renaissance astrology. Science is fallible and falsifiable, but this is not because of its reference to the world, but because the scientific institution does not always deliver results in accordance with the consensus." To quote Harry Collins, "It is not the regularity of the world that imposes itself on our senses, but the regularity of our institutions and beliefs that imposes itself on the world."[17] Let us look at some of his arguments.

Collins regards the replicability of experiments as an essential feature of natural science and as a problem for the constructivist program. He has no difficulty in showing that at every level of decision about what is a valid scientific replication, "arbitrary" decisions have to be made about how to apply any "rules" that might be suggested. He describes an "awkward student"[18] who always rejects the habits his teacher is trying to inculcate, by reinterpreting previous rules of usage in unexpected but internally consistent ways. Replicability is what scientists can be got to accept as such, for reasons other than the regularity of the world.

There are two aspects of this argument that ought to be kept separate. We can allow the less contentious Wittgensteinian point that linguistic classifications are purely conventional, reinforced as it is by accounts of radically different classifications of objects in different cultures. Moreover let us agree that there are no explicit rules for applying social conventions: they are learned as habits and acquired as skills; they are "know-how," not "know-that."

The second aspect of the argument is more contentious and more interesting. It is that the Wittgensteinian story about alternative "forms of life" cannot be used to explain why many of our socially entrenched expectations are in fact fulfilled *without* the ingenious reinterpretations of Awkward Student. Normal social life and communication would be impossible if this were not the case. Collins agrees that the existence of order is taken for granted and rightly sees that this is what requires explanation as much as does change. The Humean philosopher of induction of course does not claim to give such an explanation: he merely points to brute facts. Collins goes further—he does not even allow us any independent facts to point to and makes his explanation depend purely on social stipulation. But surely without the assumption that social habits have grown to reflect some order in the world, this cannot be enough to explain the general success of the expectations we have to rely upon for everyday practical needs. Some biological needs are common to all cultures, and would explain why many inductive assumptions are the same in all cultures in similar environments. In other respects, different cultures may have different needs and correspondingly different linguistic classifications. In neither case does it follow that all inductive regularities are purely conventional. On the contrary, insofar as their varying expectations are satisfied, it implies that objective order in the psycho-social-natural world in which we all live is more various and multifaceted than our culture recognizes—*more* various, but not *infinitely* various so that any old classification will do for any given social purposes.

There seems here to be an extraordinary neglect of the biological basis of perception, which must for reasons of survival relate to real regularities in the world. Collins accepts, admittedly as just a scientist's consensus, the current story about most of science, but it is very odd to accept this and not to accept the consensus in biology about animal and human learning for which physical and physiological mechanisms certainly play a part in determining how similarity and regularity are perceived in nature. Does Collins assume he has a "better" scientific account of learning than this, and if so how can this be, seeing that in his terms there necessarily is no consensus now about any alternative account he may give?

A second point concerns the holism of perceptual and scientific systems. Collins speaks as though replicability of experiments is an entirely independent matter that can be interpretively imposed at will, and has so allowed himself to be seduced by the positivist emphasis on "verification" of individual experimental results that he retains this atomism of experiment even in his constructivism. He neglects the point of the Duhemian conception of holism of theory, which is not that *all* individual replications can be reinterpreted at will but that *some* can, while being constrained by others, and by the coherence of the whole theoretical network. If Collins replies that the conceptions of "experiment" and "coherence" are also social conventions, that may be true, but these are conventions of a special kind, as it were, meta-conventions, which are consequences of the ideal type of science in our society. Once the ideal type is recognized, replications within its framework *cannot* be constructed entirely at will.

This brings me to a third and most important objection to constructivism of Collins' type, and that is the confusion, which is a *sociological* confusion, between science as a recognizable institution, and cognitive systems that are other than science. That there is an ideal type of science is a social fact and hence something that cannot be neglected by sociologists of science of all people. It is an ideal type that certainly includes the notion of replicability of experiments, and it also serves to demarcate "science" from "pseudo-science" in relation to the organs of the scientific establishment. It follows that case histories from pseudo-science, failed science, or not-yet-acceptable science, such as parapsychology, studies of the mental life of plants, or Blondlot's "N-rays," to which Collins and others devote considerable attention,[19] cannot be used as arguments against objective natural replicability in ideal-type science. They can, at best, show that these studies do not belong to the ideal type. It is the "normal sciences" that provide the most difficult and most mysterious cases for the constructivist thesis. Why is there order in the apparently most successful physical theories? This is an up-side-down question from the point of view of orthodox realism, because it opens up the possibility that the orthodox view may have to be modified for all science, hard as well as soft, normal as well as pseudo. Indeed it does, but I shall argue that necessary modifications in orthodoxy can take place without resorting to radical constructivism.

As examples of "normal" science, Collins describes the partially successful attempts to replicate the TEA-laser in a variety of laboratory environments, and the unsuccessful attempts to replicate Joseph Weber's "gravity waves."[20] The most noteworthy thing about these accounts is the immense trouble people went to, to *wrestle with material objects* in order to satisfy the conditions of replicability, and also the fact that some such attempts *fail*, whatever the antecedent social expectations were. Is it conceivable that such problems should arise with sheer manipulation if all questions of replicability could be settled by social fiat without reference to the world? If a constructivist replies that accepting sheer manipulation as a constraint is also a matter of convention, this is just to slip again from conventions *within* science to conventions *about* the ideal type of science itself.

How does it happen that there is eventual "closure" of scientific disputes about what "order" is to be accepted? Collins regards closure as a social decision not forced by the "facts." Not forced perhaps, but it does not follow that social decision has nothing to do with objective nature. It is worth referring here to Martin Rudwick's recent detailed study of just this problem of closure with respect to the controversy about the Devonian rocks in mid-nineteenth-century geology.[21] The geological history of the Devonian strata was disputed among experts between the years 1835 and 1850, after which there was closure of the debate which has persisted to the present time. Rudwick explicitly addresses himself to the question whether this closure was wholly due to social and professional factors favoring the opinion which became orthodoxy, or whether it did indeed reflect the objective but unobservable history of the rocks, the theory of which was eventually based on sufficient evidence. After an exhaustive inquiry into all possible social factors, he adopts the objectivist conclusion, but not for want of rigorous search for arguments for its opposite.

4. Micro- and Macro-sociology

The constructivist thesis is that scientific consensus is in principle no different from any other sort of persuasion of people to believe in a political, ideological, or religious system, or even to believe for purposes of their own class or personal or professional advancement. Science is a complex set of social institutions and one would expect that the constructivist thesis would be backed by detailed *macro*-sociological inquiry. Collins himself remarks that "the latest trends in empirical studies of contemporary science is a focus on more and more detailed and microscopic aspects of life within the laboratory . . . their narrow field makes it difficult for them to take account of the wide social base of legitimate knowledge."[22] Yet his own case studies are all micro-sociological.

Similarly, in their Introduction to the collection of papers on sociology of science *Science Observed*, Knorr-Cetina and Mulkay speak of a social methodology they call "internal."[23] This is meant not in the old sense of being exclusively concerned with self-contained technical arguments and experimental evidence, but in the sense of offering micro-studies of specific groups of scientists, relatively cut off from surrounding society. Sociologists of science go into the laboratory as into a strange tribe, and like ethnomethodologists they concern themselves with portraits of closed scientific communities with internally defined universes of meaning. Some of these studies have taken what Knorr-Cetina and Mulkay call the "linguistic turn," toward modes of scientific communication and discourse. Literary and taped sources are taken as data, and claimed to be more "objective" than the participatory impressions of the ethnomethodologist.

There is a parallel here with modes of theorizing in social anthropology. Internal ethnomethodology echoes the Durkheimian concern with symbolic religious systems seen as "a synthesis *sui generis* of particular consciousness" which "sometimes indulges in manifestations with no purpose or utility of any sort, for the mere pleasure of affirming itself."[24] And more recently in anthropology too, symbolic meaning analysis gave way to structural analysis

of discourse—reported and repeated myth rather than interpreted ritual behavior became the basic data. More recently still, this gives way to deconstruction, represented
most strikingly in historical sociology of science by Michel
Foucault.

Micro-sociology of science is generally constructivist in
its repudiation of standard philosophy of science and of
scientific realism, regarding these as misleading distor-
tions of scientific practice. There is much discussion of
scientific artifacts, public rhetoric, power struggles within
the laboratory, and between laboratories and funding
agencies, and the false consciousness said to be associated
with an ideology of truth and of positive knowledge. The
picture of science is of an embattled and incestuous community akin to the asylum or the prison. In contrast,
Foucault does release science into a wider macrosociologi-
cal world, but still without any general macro-explanation
of the institution of science as such. Indeed he would
probably disclaim the possibility of any such unified explanation based on a unified epistemology of science.

The problem with much sociology of science is that it is
very difficult to discover what the institution of science is
for. Woolgar has a somewhat sinister explanation of why
this is so.[25] He complains that the latent aim of constructivism seems to be to debunk or demystify science with
ideological purposes in mind, and believes that if this is
the case, the aim should be made explicit and pursued
positively instead of by default. But as I have indicated in
talking about the actors' model, sociology ought to look
for manifest functions as well as latent false consciousness,
revelatory criticism as well as demystification.[26]

Sociologists of science have fallen too easily into the
positivist trap and concluded that the absence of precise
rules and logical definition for scientific method implies
the futility of any categorization of science as a social
institution. Failing such understanding, we can perhaps
do no better than return for hints to the demarcation

criteria described by the positivist philosophers, but without their inappropriate demands for sharp boundaries and rules and logical rigor. One of the elements of ideal-type science must be social interest in predictability and control in the natural environment—confirmation, falsifiability, instrumentality, objective replicability, technical interest, or however it may be specified. (Incidentally Collins' Awkward Student would not get very far with this social function). This function cannot be realized in social practice unless there is some element of reference in the relation of science to the world, and this is ultimately the reason why science exhibits order. Moreover our society is interested in exploiting that order, and in creating artificial objects, processes, and closed systems on its basis. Without a basis no social explanation can be complete. Thus the generalized epistemology of science that makes it the science we know is a technical and instrumental one.[27]

No doubt, however, the social function of science goes beyond technical exploitation to an interest in providing socially legitimating myths and cosmologies. Sociologists and historians of science have amply illustrated the degree of conventionality that exists in high-level theory, as opposed to low-level instrumental application, which cannot be wholly conventional. They have suggested, for example, why there is sometimes a strong social interest in the ideology of the realism and truth of science (T. H. Huxley's battle with the Church for social and political influence, for instance).[28] They have suggested why scientific societies, unlike traditional ones, seem to have an interest in conceptual innovation and a continually changing body of positive knowledge. Apart from the desire for technical advance, new theories may undermine entrenched and undesirable power-groupings; they may have expressive and aesthetic functions; or they may merely satisfy the human desire to solve puzzles and play games, which can be indulged in a relatively affluent society. Whatever the answers to these and other questions, it is clear that they require macro- as well as micro-sociological approaches.

What, then, are the characteristics of a socialized epistemology? I have suggested four somewhat irenic points, where apparently conflicting philosophical approaches may be reconciled within a generally socialized framework:

1. Where a system of knowledge or cognitive belief is either accepted (as in our own science) or claimed to be cognitive in other cultures or by subgroups in our own culture, there can be internal epistemological study of the relations claimed to hold between data, theory and conceptual frameworks, and external explanation of the genesis of concepts and methods, and of the goals and interests served by the cognitive system. These tasks depend on the fact that any social system is held together by some set of social rules or norms, and can be reconstructed (albeit contestably) by an ideal social type.

2. "Cognitive systems" can be defined sufficiently widely to capture all symbolic systems or "models of the world" that are exhibited in ritual and/or developed for technical and social purposes.

3. A whole range of sociological models is available for the explanation of natural science and has been exploited in recent studies: from the micro-sociological point of view, the actors' model, rational action, ethnomethodology; from the macro-sociological point of view, various types of symbolic action, functionalism, evolutionism, structuralism, deconstruction. There is no single "correct" model, only models more or less useful and illuminating in particular cases.

4. There is a variety of viable epistemologies of science, other than the extremes of the two dimensions realism-relativism, rationalism-constructivism. The emphases of different moderate epistemologies are not inconsistent, but subject to sociological and historical testing like any other methodological hypothesis. What is true about ac-

counts of scientific rationality can be accommodated within this general scheme, but without allowing it unique access to the cognitive—that is, in a more traditional sense, to the rational.

NOTES

1. W. V. O. Quine, "Epistemology Naturalized," in *Ontological Relativity* (New York: Columbia University Press, 1969), pp. 69–90; see p. 69.

2. E. Durkheim and M. Mauss, *Primitive Classification* (1903), trans. Rodney Needham (London: Cohen & West, 1963); E. Durkheim, *Elementary Forms of Religious Life* (1912), trans. Joseph W. Swain (London: Allen & Unwin, 1915).

3. See, for example, Durkheim's *The Rules of Sociological Method* (1985), trans. Sarah Solovay and John Mueller (Chicago: University of Chicago Press, 1938). 'Positivism' will be used in this paper to include all methodologies that maintain the objective, value-neutral stance of scientific explanations, and adopt criteria of empirical tests as decisive for rational knowledge of the world.

4. See Douglas' "grid and group" theory of styles of cognitive thought in her *Natural Symbols* (New York: Pantheon Books, 1970), chap 4; and Bloor's *Knowledge and Social Imagery* (London: Routledge & Kegan Paul, 1976).

5. M. Hesse, "The Strong Thesis of Sociology of Science," in *Revolutions and Reconstructions in the Philosophy of Science* (Bloomington: Indiana University Press, 1980), chap. 2.

6. See the discussion in B. Barnes, *Interests and the Growth of Knowledge* (London: Routledge and Kegan Paul, 1977), chap. 1. See also the "weak program" of D. E. Chubin and S. Restivo, "The 'Mooting' of Science Studies: Research Programmes and Science Policy," in *Science Observed,* ed. K. D. Knorr-Cetina and M. Mulkay (London: Sage Publications, 1983), pp. 53–84; see pp. 60, 76, which does not presuppose that "scientific method" is the best way to study science itself.

7. P. Forman, "Weimar Culture, Causality and Quantum Theory, 1918–1927," *Historical Studies in the Physical Sciences* 3 (1971): 1–115. B. Wynne, "Physics and Psychics: Science, Symbolic Action and Social Order in Late Victorian England," in *Natural Order,* eds. B. Barnes and S. Shapin (Beverly Hills, Calif: Sage Publications, 1979); pp. 167–84; see p. 167.

8. J. Hendry, "Weimar Culture and Quantum Causality," *History of Science* 18 (1980); 155–180.

9. H. M. Collins, "Stages in the Empirical Programme of Relativism," *Social Studies of Science* 11 (1981): 3–10.

10. For example, B. Wynne, "Physics and Psychics," and S. Shapin, "Homo Phrenologicus: Anthropological Perspectives on an Historical Problem," in *Natural Order*, p. 41.

11. R. Needham, *Belief, Thinking and Experience* (Oxford: Blackwell, 1972).

12. See for example, P. Pettit and G. MacDonald, *Semantics and Social Science* (London: Routledge & Kegan Paul, 1981), chap. 1: The rational action model goes back to Weber's discussion of *Zweckrationalität*, but unlike the positivists Weber supplemented this purposive action by a concept of *Wertrationalität*, that is, rational action recognized to be of value for its own sake.

13. E. Leach, *Political Systems of Highland Burma* (Cambridge, Mass.: Harvard University Press, 1954), p. 14.

14. B. Wynne, "Physics and Psychics," pp. 168, 184.

15. R. Horton, "Tradition and Modernity Revisited," in *Rationality and Relativism*, ed. M. Hollis and S. Lukes (Cambridge, MIT Press, 1982), p. 201; and E. Gellner, *Legitimation of Belief* (London: Cambridge University Press, 1974), p. 157.

16. This approach is developed by M. Arbib and M. Hesse in *The Construction of Reality* (New York: Cambridge University Press, 1986).

17. H. M. Collins, *Changing Order* (London: Sage Publications, 1985), p. 148. For a more detailed critique of Collins' constructivism, see my "Changing Concepts and Stable Order," *Social Studies of Science*, forthcoming.

18. *Changing Order*, p. 13.

19. Ibid., p. 45 and chap. 5.

20. Ibid., chaps. 3 and 4.

21. M. Rudwick, *The Great Devonian Controversy* (Chicago: University of Chicago Press, 1985), chap 16.

22. *Changing Order*, p. 153, n. 7.

23. Knorr-Cetina and Mulkay, eds., *Science Observed*, p. 6.

24. Durkheim, *Elementary Forms of Religious Life*, p. 423.

25. S. Woolgar, "Irony in the Social Study of Science," in *Science Observed*, 239–66; see pp. 254, 261.

26. B. Latour has a subtle illustration of manifest functions which are not entirely instrumental in his "Give me a laboratory and I will raise the world," in *Science Observed*, p. 141. He shows how Pasteur deliberately united the traditional methodology of the laboratory and applications on the farm in his work on anthrax. He effectively *con-*

structed anthrax as a scientifically and publicly recognized problem, whereas before it had been no more than a fact of life—a social plague about which no problem arose because in the vast variety of circumstances where "anthrax" struck, nothing could be done. But there are also implicit presuppositions of realism, since the incidence of a number of killing and uncontrollable diseases going under the name of "anthrax" was eventually reduced and controlled.

27. For more on modern debates about "realism and instrumentalism" see my "What is the point of pointer-readings?" prepared for the Workshop on "Contexts of Dialogue for the Formation of Quantitative Concepts in Science," Tel Aviv and Jerusalem, 1986.

28. See J. Brooke, "Science and the Fortunes of Natural Theology: Some Historical Perspectives," forthcoming.

SOME DIFFERENT CONCEPTIONS OF
RATIONALITY

Richard Foley

1. A Characterization of Rationality

Contemporary philosophy is filled with discussions of rationality. Epistemologists and philosophers of science have been concerned with what an individual must *believe* if that individual is to be rational, while ethicists and decision theorists have been concerned with what an individual must *do* to be rational. However, proponents of both kinds of accounts, accounts of rational belief and accounts of rational actions, have had relatively little to say about the concept of rationality itself. They have been primarily concerned to propose standards of rational belief or rational action, but little has been done by way of offering a general characterization of rationality.

This has been unfortunate, since without some general characterization, it is difficult to adjudicate among various accounts of rationality. Consider contemporary epistemology, for example. Prominent contemporary epistemologists have proposed radically different accounts of rational belief. According to some, an individual's belief is

This paper is a revised version of "Epistemic Rationality and Scientific Rationality," which appeared in *International Studies in the Philosophy of Science: The Dubrovnik Papers* 1(1987), 233–50 and appears here with permission of the editors of that journal.

rational just in case the individual has (or can generate) something like an adequate argument for the belief. To be sure, there are significant disagreements among these epistemologists about what constitutes an adequate argument. Foundationalists insist that the argument must have premises that are properly basic for the individual, where this means (roughly) that it is rational for the individual to believe each of the premises because each premise constitutes an argument for itself; each premise is self-evident. By way of contrast, coherentists deny that there are any properly basic claims. They insist that an individual's beliefs are rational just in case they are appropriately coherent, where this means (again roughly) that every belief is such that the individual's remaining beliefs constitute a good argument for it.

The differences between foundationalists and coherentists are significant, but they are minor in comparison with the differences between foundationalists and coherentists, on the one hand, and proponents of reliabilism, on the other. Reliabilists reject the idea that an individual must have or be capable of generating an adequate argument for what he or she believes if the belief is to be rational. Rather, they say (once again, roughly) that an individual's belief is rational just if the process that produced the belief, whether this be a perceptual process or a memory process or a reasoning process or whatever, is reliable. Then there are various proposals for understanding reliability. A reliable process might be said to be one that is likely to generate true beliefs and fail to generate false beliefs in the long run, or it might be said to be one that would generate true beliefs rather than false beliefs in most close counterfactual situations. Or it might be said to be yet something else.[1]

But whatever the precise nature of reliability is taken to be, reliabilist accounts of rational belief are radically different from those proposed by foundationalists and coherentists. Foundationalists and coherentists, despite their differences, both endorse accounts of rational belief that emphasize the perspective of the individual believer. The emphasis is upon what the individual believes, what experiences he or she has had, what he or she remembers, etc., and whether in terms of these things he or she has good arguments for various propositions. Reliabilists, by contrast, endorse accounts of rational belief that deemphasize the perspective of the individual believer.

Matters become even more complicated when one turns away from epistemology proper and considers other areas of philosophy that are concerned with questions of rational belief. Consider philosophy of science, for instance. Philosophers of science often propose standards of rational belief (or rational hypothesis acceptance) that are significantly different from the standards proposed either by foundationalists and coherentists, on the one hand, or by reliabilists, on the other hand. For example, they often propose that it is rational, all else being equal, to prefer the simpler of two hypotheses, and likewise they often propose that it is rational, all else being equal, to prefer the more fertile of two hypotheses. But among traditional foundationalists, coherentists, and reliabilists, the virtues of simplicity and fertility are only rarely mentioned, much less advocated as criteria of rational acceptance.

It would be possible to describe yet other philosophical approaches to questions about rational belief, but to anyone familiar with contemporary philosophy the point here should be obvious: there are a variety of different kinds of accounts of rational belief that are advocated by contemporary philosophers and these accounts often seem to have little contact with one another. Of course, it is not unusual for philosophers to disagree, but the nature of the disagreement here is unusual. It is as if the proponents of these accounts are not even concerned with the issues that motivate the proponents of the other accounts. It is as if the proponents of the various accounts are trying to answer different questions. This makes especially pressing the problem of what is to be made of the variety of such accounts. Which of these very different kinds of account of rational belief is the most defensible? Unfortunately, this question is one that cannot even be profitably discussed until something is said about the nature of rationality. Without a general characterization of rationality, there will be no criteria by which to judge one of the approaches more plausible than the other. All there will be to go on are "intuitions" about what it is rational for an individual to believe in various situations.

How then might rationality be characterized? As a way of taking a first step toward a general characterization of rationality, I will suggest two theses about the nature of rationality, theses that have far-reaching implications. The first is that rationality is essentially goal-oriented and that, correspondingly, claims about the rationality (or irrationality) of an individual's actions or beliefs are at bottom claims about how effectively the individual is pursuing those goals.[2] However, there is an immediate complication. Claims about how effectively an individual is pursuing his or her goals can be made from a variety of perspectives. They can be made from the perspective of the individual or from the perspective of the community or from the perspective of most scientists or from any number of other perspectives. Moreover, these claims can differ depending on the perspective from which the claim is made. From the perspective of the individual himself, some course of action might seem to be an obviously effective means to his goals. However, from the perspective of most people in the individual's community or from the perspective of most scientists or from the perspective of most experts, this very same course of action might seem to be an obviously ineffective means to the individual's goals.

This complication gives rise to a second major thesis about the nature of rationality. Namely, no single perspective is *the* privileged perspective from which assessments of rationality are to be made. Claims of rationality properly can be made and are made from a number of perspectives. To illustrate this, consider three general schemas of rationality (although theoretically there perhaps is no limit to the number of schemas that might be introduced):

1. The Aristotelean schema: All else being equal, it is rational for an individual S to bring about Y if he has a goal X and if on careful reflection he would believe that Y is likely to be an effective means to X.

2. The radically subjective schema: All else being equal, it is rational for an individual S to bring about Y if he has a goal X and if he believes that Y is likely to be an effective means to X.

3. The radically objective schema: All else being equal, it is rational for an individual S to bring about Y if he has a goal X and if Y in fact is likely to be an effective means to X.

None of these schemas is clearly counterintuitive. On the contrary, each when thought about in a certain way can seem to be the obviously correct schema. Consider a simple example. Suppose S is betting on a horse race in which only three horses are entered, and suppose further that he must bet on one of the horses to win and that the odds on each of the horses is the same. On which horse is it rational for him to bet, assuming that his goal is to win the bet?

The answer seems to depend on what elements in the betting situation one emphasizes. Suppose that S has a good deal of information about horse A in comparison with the other two horses, information about A's past record, his workout times, the jockey who is riding him, and so on. And suppose that were S to reflect upon this information, he himself would acknowledge that, given this information, it is highly likely that A will win the race. Perhaps A's record is better than the records of the other two horses, and perhaps his recent workout times are also better. Moreover, perhaps the best jockey at the track is riding him. Then, S would seem to have a good reason to bet on A. Indeed, were he to be reflective, he himself would be critical of any other bet; he himself would think it a mistake to bet on any other horse. And so, in accordance with schema 1 above, it would seem to be rational for him to bet on horse A.

However, suppose that S does not reflect carefully upon the information that he has about horse A. As a result, he believes that horse B will win. In such a case, we can be critical of S for not attending to the evidence more carefully. Even so, given that he believes B will win, in at least one sense he would be acting irrationally if he betted on either of the other horses. For a decision to bet on one of the other horses would amount to a decision to bet on a horse that he believes will lose. And this would constitute a paradigm of irrationality, given that his goal is to win the bet.

But now, suppose that although S has good evidence for thinking that A will win and although S believes B will win, in fact it is highly likely that C will win. Suppose that unbeknownst to S, both horse A and horse B have injuries that will prevent them from performing well. As a result, only C is healthy. Then doesn't S have a reason to bet on C? After all, this is the bet that actually is likely to succeed in getting him what he wants—that is, to win the bet. And so, in accordance with schema 3, it would seem to be rational to bet on C.

Which of the three bets is *the* rational one for S to make here? And which general schema of rationality is *the* correct one? I do not think there is an answer to these questions. Each schema is plausible. Each simply reflects a different perspective from which evaluations can be made of how effectively S is pursuing his goals. And as the above example illustrates, depending upon our purposes, interest, etc., any one of these perspectives can seem to be the appropriate one for making such evaluations. If, for example, S bets on horse B, schema 2. allows us to see the reasonability of his decision, given his present beliefs— i.e., given his present (unreflective) perspective. However, by hypothesis S himself would be critical of this decision were he to be reflective. So, if we want to give expression to the sense in which S has violated even his own standards, we will be drawn to schema 1. On the other hand, if we realize that neither S's actual decision nor a decision in accordance with his own standards is likely to accomplish what S wants such a decision to accomplish and we want to give expression to this sense in which he has a reason not to make either decision, we will be drawn to schema 3.

What this simple example suggests is that claims of rationality tend to be elliptical. They are made from a variety of perspectives, but often the perspective is not made explicit. If we claim that an action or a belief of an individual is rational, we may be claiming that from the individual's own unreflective perspective the action or belief seems to be an effective means to the individual's goals. Or we may be claiming that from what would be the individual's perspective were he or she to be carefully reflective, it seems to be an effective means to these goals. Or we may be claiming that from yet some other perspective (the perspective of most members of the individual's community or the perspective of most scientists or the perspective of experts or even, in the limiting case, the perspective of an ideal observer), it seems to be an effective means to these goals.

So, claims of rationality tend to be elliptical. They tend not to make fully explicit the perspective from which the evaluation is being made. Suppose, then, that we make fully explicit the perspective from which we are making our judgments about the rationality of an individual's beliefs. Does this solve the problem? Not quite, for there is a second way in which claims of rationality tend to be elliptical. They tend to be elliptical with respect to which of an individual's goals are to be taken into consideration. Are the individual's beliefs or actions to be judged against the standard of how effectively they promote the total constellation of goals, or are we to single out a certain subset of these goals and then evaluate the individual's actions or beliefs against the standard of how well they promote the satisfaction of the goals in this subset? And if the latter, which of the individual's goals belong in this subset—i.e., which are relevant for an assessment of the individual's actions or beliefs?

In trying to answer these questions, it is tempting to claim that the goals relevant to an assessment of an individual's beliefs are in principle different from the goals relevant to an assessment of an individual's actions. In particular, it is tempting to group together all of an individual's goals that involve his or her own well-being and comfort, the well-being and comfort of others, and other like concerns under the heading of practical goals and then to use these goals to evaluate the rationality of the individual's *actions.* An individual's actions are rational, then, just in case (given some agreed-upon perspective) these actions seem to do the best job (or at least an acceptably good job) of satisfying these practical goals. It is equally tempting to group together all of an individual's goals that concern his or her coming to understand the world under the heading of intellectual (or theoretical) goals and then to use these goals to evaluate the rationality of the individual's *beliefs.* An individual's beliefs are rational, then, just in case (given some agreed-upon perspective) these beliefs seem to do the best job (or at least an acceptably good job) of satisfying these intellectual goals.

Although this way of distinguishing the rationality of actions from the rationality of beliefs may be tempting, there is little to recommend it. After all, how one *acts* can dramatically affect one's intellectual prospects, and insofar as this is so it is irrational, all things considered (i.e., when all of one's goals are taken into account), to discount one's intellectual goals when deciding what to do. It perhaps is not so obvious but it is equally true that the reverse also can be the case. Namely, what one believes can dramatically affect one's practical prospects, and insofar as this is so it is equally irrational, all things considered, to discount one's practical goals when deciding what to believe.[3]

Thus, there is no *a priori* reason to discount intellectual goals in determining what it is rational to do, and likewise there is no *a priori* reason to discount practical goals in determining what it is rational to believe. This is not to say, however, that a claim about the rationality of an individual's actions or beliefs must be a claim about how effectively those actions or beliefs satisfy the total constellation of the individual's goals. It is not to say, in other words, that a claim about the rationality of an individual's actions or beliefs can only be a claim about what it is rational for an individual to do or to believe, all things considered—i.e., when all of the individual's goals, both practical and intellectual, are taken into account. On the contrary, it often is both useful and altogether appropriate to make a claim about the rationality of an individual's actions or beliefs with respect only to a stipulated subset of an individual's goals. Thus, for example, although there is no *a priori* reason why we must discount practical goals in deciding what it is rational for an individual to believe, as a matter of fact it often is both useful and altogether appropriate to do so. That is, we can ask what an individual has reasons to believe *insofar* as he or she has such-and-such intellectual goals. In making such evaluations, we are in effect making a judgment about what it *would* be rational, *all things considered,* for the individual to believe *were* these intellectual goals his or her only goals.

Thus, claims of rationality can be elliptical with respect to the goal that is being presupposed, as well as with respect to the perspective from which they are being made. All claims of rationality are goal-oriented; they are claims about how effectively the individual, via his beliefs or his actions, is pursuing his goals. But, not all of the individual's goals need be taken into account in evaluating the rationality of an individual's actions and beliefs. Sometimes we are interested in evaluating the rationality of an individual's actions or beliefs insofar as he or she has goals of a certain kind (and ignoring any other goals the individual might have).

More generally, we can say that implicit in every claim of rationality is a *point of view,* where a point of view can be thought of as consisting of a goal (or goals) and a perspective. Thus, every claim about the rationality of an individual's beliefs or actions is (at least implicitly) a claim made from a certain perspective (say, the individual's own perspective, on reflection, or the perspective of most members of the community) concerning how effectively the individual's beliefs or actions satisfy this goal (say, a prudential goal or an intellectual goal or perhaps the total constellation of goals).

2. Foundationalism, Coherentism, and Reliabilism

Thinking about claims of rationality in this way gives rise to some surprising conclusions. Most notably, it gives rise to the idea that seemingly rival accounts of rational belief or rational action may not always be genuine rivals at all. Rather, they may be accounts that are most charitably interpreted as accounts that have different subject matters, sometimes only subtly different but sometimes startlingly different. Let me illustrate what I have in mind by returning to the question of which of the two approaches to epistemology described above, the one advocated by coherentists and foundationalists and the other by reliabilists, is more plausible? This question, which is a much debated one in contemporary epistemology, assumes that the two groups of epistemologists are proposing rival accounts of rational belief. However, if we think about claims of rationality in the way I propose, as presupposing a point of view, this is no longer obvious. Given my characterization of the nature of rationality, foundationalists and coherentists are perhaps best construed as being concerned with evaluating from an Aristotelian perspective (where one adopts the individual's own reflective perspective) how effectively the individual is pursuing a present-tense intellectual goal, the goal of now believing propositions that are true and now not believing propositions that are false.[4] So, the concern here is not with what it is rational for an individual to believe insofar as he or she has the goal of having true beliefs and not having false beliefs in a few years, or in a few months, or even in a few moments; the goal is now to believe those propositions that are true and now not to believe those propositions that are false. And the concern is to evaluate how well, on reflection, the individual is pursuing this goal from that individual's own perspective.

Reliabilists, by contrast, are not best construed as being interested in evaluating individuals from this point of view. Those reliabilists who say that a reliable process is one that is likely to generate true beliefs and not generate false beliefs in the long run are best construed as being interested (or perhaps primarily interested) in a future-tense intellectual goal, that of having true beliefs and not having false beliefs in the long run. Those who say that a reliable process is one that is likely to generate true beliefs and not to generate false beliefs in close counterfactual situations can be construed as being interested in the same present-tense intellectual goal that interests foundationalists and coherentists, the goal of now having true beliefs and now not having false beliefs. But on the other hand, each kind of reliabilist is interested in evaluating how effectively an individual is pursuing the goal in question, whether it be the long-term goal or the present-tense goal, not from an Aristotelian perspective but rather from a more objective perspective, the perspective of an external observer who knows which belief acquisition processes are likely to generate truths in the long run or in close counterfactual situations—hence, the requirement

that a belief is rational only if it is the product of a process that is reliable in the required sense.

One way of illustrating the difference between the point
of view adopted by foundationalists and coherentists and
the point of view adopted by reliabilists is to imagine
situations in which an individual could increase both his
present and long-term chances of believing truths if he
now were to believe something that on reflection he would
take to be false and perhaps even obviously false. The
most dramatic of these situations are those involving evil
demons and the like. In the most general terms, the foundationalist and coherentist response to such situations will
take one of two forms. The first is a Cartesian kind of
response. It consists in arguing that such situations, at
least as typically described, are impossible; we could not
have the beliefs, experiences, and so forth that we do in
fact have and yet be under the control of an evil demon
who insures that most of these beliefs are false.[5] The second kind of response, which is more important for my
purposes here, allows that such situations may well be
possible but denies that they have any straightforward implications for what it is rational for us to believe. A situation in which we have the same beliefs, experiences, etc.,
that we now have but in which we are under the control of
an evil demon may well be one in which we are prevented
from having knowledge (since we are prevented from having true beliefs), but it is not one in which we are also
prevented from having rational beliefs. For in such a situation, we would from our perspective have the same arguments for our beliefs as we now have for them. So, insofar
as such beliefs are rational for us in what we take to be the
actual situation (with no demons), they would also be rational for us in demon situations.

By contrast, reliabilists will be inclined to say that in a
demon situation these beliefs would not be rational, since
by hypothesis the processes that produced the beliefs
would not be reliable.[6] On the other hand, were we to

have certain other beliefs, beliefs that on reflection we would not be inclined to defend, they might very well be rational. For example, imagine that the demon unbeknownst to us has arranged the world in such a way that believing the first thing that pops into our heads is a highly reliable (and perhaps also the *only* reliable) method of belief acquisition available to us. So, were we in general to believe the first thing that pops into our head, our present and future prospects for believing truths and not believing falsehoods would be significantly enhanced.[7] Now, consider a proposition *P* that is the first answer to a certain question that has popped into *S*'s head. However, *P* is such that were *S* to deliberate even for a moment about its truth, he would be convinced that it is false. Indeed, perhaps a moment's reflection would convince him that it is *obviously* false, and further reflection would not change his mind. Even so, from the point of view adopted by reliabilists, believing *P* might very well be rational, since from the perspective of an external observer who is able to distinguish reliable processes from nonreliable processes, it will seem that believing the first thing that pops into his head is a good way for *S* in this world to satisfy both his present and long-term intellectual goals. From this point of view, *S* has a reason to believe the first thing that pops into his head because this is in fact the most effective strategy for him to use in pursuing his goal of believing truths. But from the point of view adopted by foundationalists and coherentists, believing *P* is apt to be irrational, since from the perspective of the individual himself on reflection it will seem that believing *P* is likely to frustrate the goal of now believing true propositions and now not believing false propositions.

The lesson, then, is this: Perhaps the best way, where by this I mean the most charitable way, to interpret the positions of foundationalists and coherentists, on the one hand, and reliabilists, on the other, is in terms of distinct points of view. But then, the two kinds of accounts, contrary to prevailing opinion, are not genuine rivals at all. Rather, they are concerned with different kinds of rational belief. Accounts of the former kind propose standards for evaluating the beliefs of an individual with respect to a present-tense intellectual goal and from the perspective of the individual himself. Some reliabilists' accounts propose standards for evaluating the belief of an individual with respect to a future-tense intellectual goal and others with respect to a present-tense intellectual goal, but either kind of reliabilist account adopts an external perspective. Accordingly, from the point of view adopted by reliabilists there might be something to be said in favor of an individual's belief, while from the point of view adopted by traditional epistemologists there might be something to be said against that very same belief. Moreover, the former gives expression to what is favorable about the belief by saying that it is rational, while the latter gives expression to what is unfavorable about the belief by saying that it is irrational. Nevertheless, given the elliptical nature of claims of rationality, there need be no genuine conflict here. Each might be right.

3. Scientific Rationality

Thinking about rationality in this way can also illuminate the work of other contemporary philosophers who are interested in questions of rationality. To take but one example (there are many), consider recent writing in the philosophy of science on the notion of rational belief. As I mentioned earlier, one striking characteristic of this work is how dissimilar the standards of rational belief proposed by philosophers of science tend to be from the standards of rational belief proposed by traditional epistemologists. There probably is no single, simple explanation of this puzzling phenomenon, but a partial explanation might be that the two groups are concerned with different kinds of rational belief. Traditional epistemologists, I have suggested, are best understood as adopting a point of view that consists of the individual believer's own perspective (on reflection) and a synchronic intellectual goal, a goal that is concerned only with the present time-slice. But of course, most of us have diachronic goals as well as synchronic goals. We want to do well not just at the present moment. We also want to do well over time. And, this applies to our intellectual goals as well as to our prudential goals. In addition to the goal of now having true beliefs and now not having false beliefs, most of us have the goal of having true beliefs and not having false beliefs at future times as well. Moreover, in evaluating how well an individual S is pursuing this diachronic goal, it is possible to adopt an intersubjective perspective rather than a straightforwardly objective perspective or a straightforwardly subjective perspective. In particular, it is possible to adopt the perspective of the individual's community. This need not mean that we imagine the community taking a vote concerning what methods of belief acquisition are likely to promote our having true beliefs and not having false beliefs over the long run. The standards of evaluation might be determined by the community indirectly. For example, S's beliefs might be evaluated by reference to the intellectual standards of those commonly recognized by S's community as experts.

So, suppose we adopt such a point of view, a point of view consisting of this diachronic goal and the perspective of those commonly recognized by the community as experts. And suppose these experts identify some method of belief acquisition M (or some collection of methods M) as the method that it is best to use in achieving this goal. Then, all else being equal, it is rational in this community-based sense for an individual S to use this method M to acquire beliefs. Or if it is not rational for S himself to employ this method (because, say, there is a division of intellectual labor in the community), it is at

least rational, all else being equal, for him to believe the
results of those investigations on the part of experts using
M. On these matters it is rational, all else being equal, to
defer to the experts. Suppose, finally, that we add to this
sketch the fact that in our community (where "our community" is construed loosely, so that it includes, say, most
twentieth-century adults in the so-called "developed coun-
tries") it generally is scientists who are recognized as experts on intellectual matters and that most scientists (not
surprisingly) think that the collection of testing and verification procedures that constitutes the scientific method
maximizes our chances of discovering truths and avoiding
falsehoods over the long run. The result then is a gloss on
scientific rationality. Expressed most crudely, the gloss implies that claims of scientific rationality are claims about
what it is rational for us to believe, where these claims are
made from a point of view that consists (or, at least, primarily consists—I will add some qualifications shortly) of
a long-term intellectual goal[8] and the perspective of those
individuals recognized by the community as experts, the
presupposition being that in our community it is scientists who are generally regarded as experts. Given this
point of view, it is rational for an individual to employ the
methods of belief-acquisition (i.e., methods of evidence-
gathering, testing, reasoning, and so on) that would meet
the approval of the experts, where as a matter of fact the
methods approved by most experts are those implicit in
what is loosely called "the scientific method." Or short of
actually employing these methods himself, it is rational
for an individual to defer to the experts who do use these
methods.

However, as might be expected, there are a number of
considerations that complicate this relatively simple picture of scientific rationality. Notice, for one, that the long-
term intellectual goal of believing truths and not believing
falsehoods can be constrained by other kinds of goals, and
among these goals can be practical goals. Specifically, this

long-term intellectual goal might be constrained by an interest in discovering truths that are *useful*. The idea then would be to find a method of belief acquisition that is likely to yield not only truths *per se* but, in addition, truths that are easy to use in our attempts to control and to understand the world. The emphasis here is still on the discovery of truths and the avoidance of falsehoods, but the goal is no longer a purely intellectual goal. It is tinged by the pragmatic constraint of finding truths that can be made to "work" for us. The result is a bias against truths that are so complicated or so insignificant that we cannot make use of them.

The research in most areas of science seems to presuppose an interest in a complex goal of this sort. Accordingly, a promising suggestion concerning how to understand talk of scientific rationality would be in terms of such a goal. Given this suggestion, the scientific point of view (or at least one scientific point of view—there may be more than one) would differ from the point of view implicitly adopted by traditional epistemologists not only by de-emphasizing the individual perspective in favor of a community-based perspective (e.g., the perspective of most members of the community, or the perspective of those recognized by the community as experts) and not only by being concerned with a diachronic goal (where what is of interest is to find a method that eventually will lead one to truths) but also by being concerned with a practical goal (where what is of interest is to find easily utilizable truths). Expressed most starkly, the contrast is this: traditional epistemologists are best understood as being concerned with a purely intellectual goal; they are concerned to answer the question of what it is rational for an individual to believe insofar as he is interested *solely* in now having an accurate picture of the world. On the other hand, philosophers of science are concerned primarily to answer the question of what it is rational for an individual to believe insofar as he is part of a community that is interested in having (at least in the long run) an accurate *and* easily utilizable picture of the world.[9]

This quick sketch of scientific rationality is still far too simple; it leaves many questions unanswered.[10] The idea here is only to provide a framework for thinking about scientific rationality and for understanding the differences between it and the kind of rationality that typically is the concern of traditional epistemologists. Nevertheless, even from this general framework, it is possible to discern some significant consequences concerning these differences. Consider, for example, some of those characteristics that philosophers of science are prone to regard as virtues of theories—for example, simplicity and fertility. A number of philosophers of science have claimed that it is rational, all else being equal, for an individual to accept the simpler of two rival theories. Similarly, many have claimed that it is rational, all else being equal, to accept the more fertile of two rival theories.

From a purely epistemological point of view, which emphasizes the goal of now believing truths and not believing falsehoods, it is hard to see what it is that necessarily recommends simplicity and fertility. However, any conclusions here need to be at least somewhat tentative, since it is notoriously difficult to get agreement concerning exactly what simplicity and fertility are. Indeed, some have even proposed that simplicity be defined in explicitly epistemic terms—for example, in terms of how probable it is that a theory is true.[11] It then trivially will be the case that the simpler of two theories, all else being equal, is to be epistemically preferred. But short of defining simplicity or fertility in epistemic terms, it is hard to see what recommends them from an epistemic point of view. It is hard to see what there is epistemically to recommend simple theories if they are understood roughly to be theories that account for diverse phenomena in a brief unified way.[12] And it is equally hard to understand what there is epistemically to recommend fertile theories if they are understood, again roughly, as theories that have a wide scope for future development.[13] After all, one might ask, is the world such that simpler or more fertile theories are likely to be true? And even if (as is doubtful), the answer is thought to be "yes," one might still wonder whether simplicity and fertility are, as philosophers of science often seem to suggest, *fundamental virtues* of theories, which *anyone* properly can use *in any situation* to help determine whether it is rational to accept a theory, or whether they instead are *derivative virtues,* which derive from the fact (if it is one) that *we in our present* situation have independent evidence (for example, from the history of science) indicating that the simpler or more fertile of two theories is more likely to be true.[14]

So, from a purely epistemological point of view, it is doubtful whether there is as much to be said for simplicity and fertility as philosophers of science commonly do say for them. Why, then, have so many philosophers of science taken it as uncontroversial that these are fundamental virtues of theories? My suggestion is that philosophers of science often take this to be uncontroversial because they are concerned, as they should be, to explicate the kind of rationality implicit in the practice of science, and because it is best to explicate this kind of rationality from a point of view other than a purely epistemic point of view. Rather, the kind of rationality implicit in science is best explicated in terms of a point of view that emphasizes not only a diachronic intellectual goal, where the concern is to find a method of belief acquisition that will maximize our chances of discovering truths and avoiding falsehoods over the long run, but also a prudential goal, where the concern is to discover truths that are useful to us. Insofar as a concern with such goals is characteristic of the scientific point of view, there is a straightforward explanation as to why it is scientifically rational, all else being equal, to prefer the simpler or the more fertile of two theories. From a scientific point of view, it is rational,

all else being equal, to prefer the simpler of two theories
for just the reason that intuitively we would expect.
Namely, the simpler of two theories is easier to *use* (even
if, all else being equal, it is not more likely to be true). In
other words, it is the pragmatic aspect of the scientific
goal that explains why simplicity is a virtue that it is rational for scientists to cultivate in their theories. Analo-
gously, the diachronic aspect of the scientific goal explains
why fertility is a virtue that it is rational for scientists to
cultivate in their theories. For, the theory that is the more
likely to generate promising research projects is more
likely, we think, to lead us eventually toward the truth
(even if, all else being equal, it itself is not more likely to
be true).

Thus, once we accept the idea that there are these differences between the scientific point of view and the epistemic point of view (and corresponding differences
between scientific rationality and epistemic rationality),
we have a straightforward explanation of what otherwise
might seem to be the puzzling phenomenon of philosophers of science and epistemologists being disposed to
put forward very different criteria of rational belief, with
the latter commonly thinking that some of the criteria
proposed by the former are not even starters; they are not
even particularly good candidates to be criteria of rational
belief. This phenomenon is a straightforward consequence
of the fact that scientific rationality is a different kind of
rationality from epistemic rationality. Accordingly, we
should *expect* the criteria of scientific rationality to be significantly different from the criteria of epistemic rationality. Likewise, we should not be surprised if there turn out
to be situations in which it is *epistemically rational* for an
individual S to withhold judgment on a theory *T*, since the
purely intellectual evidence in its favor may be such that
believing it is an unacceptable epistemic risk, but in
which nonetheless it is *scientifically rational* for members of
S's community to accept *T*, given that there is consider-

able (albeit not definitive) intellectual evidence in its favor and given its simplicity, fertility, and other virtues.

Is this an acceptable way to leave the matter? Is it acceptable simply to claim that in such a situation it is *scientifically rational* for S to believe T but *epistemically rational* for him to withhold judgment on T and then to leave the matter at that? Must not there be some way of rationally deciding which recommendation, the recommendation of the philosopher of science or that of the epistemologist, is to take precedence? And more generally, isn't it natural to want a criterion of rational belief that tells us what is *the* rational attitude for an individual S to take toward theory T?

It *is* natural to want such a criterion, but there are a number of pitfalls to be avoided in trying to satisfy this want. Notice first that neither the theory of epistemic rationality nor the theory of scientific rationality can possibly provide us with a criterion for determining what is *the* rational attitude for an individual to take toward a theory T. The former by hypothesis is concerned only with what it is rational for an individual to believe insofar as he is interested in satisfying his epistemic goal, the goal of now believing truths and now not believing falsehoods. The latter by hypothesis is concerned only with what it is rational for an individual to believe insofar as he is interested in acquiring beliefs via a method that in the long run is likely to generate true and easily utilizable beliefs. Thus, both the epistemic point of view and the scientific point of view are restricted points of view. By this I mean that neither point of view is concerned with the evaluation of an individual's beliefs with respect to how effectively they promote the full range of an individual's goals. Rather, they are concerned only with a specified subset of the individual's goals. Accordingly, neither the recommendations of scientific rationality nor the recommendations of epistemic rationality can plausibly be regarded as recommendations about what is *the* rational attitude for S to take toward a theory *T.* Correspondingly, neither kind of recommendation can be said to take preference over the other when the two conflict. Indeed, since the evaluations are made from different points of view, it is, as I have said, misleading to say that their evaluations genuinely conflict at all (any more than, say, the evaluation of an individual as athletically talented genuinely conflicts with an evaluation of that same individual as intellectually untalented).

This is not to say, however, that it is altogether inappropriate to make claims about what is *the* rational attitude for *S* to take toward a theory *T.* On the contrary, such claims can be appropriate, subject to two important qualifications. The first is that in order to determine what is *the* rational attitude for *S* to take toward theory *T,* one must take into account all of *S*'s goals. We cannot get an answer to the question of what is *the* rational attitude for *S* to take toward *T* by identifying, as do epistemologists and philosophers of science, only some of the individual's goals as relevant. Rather, we must take into account all of *S*'s goals, asking what attitude (believing, disbelieving, or withholding) does the best overall job of promoting these goals, where this is a function, roughly, of the relative importance of *S*'s various goals and the relative likelihoods of his various options for bringing about these goals. The attitude, then, that does the best overall job of satisfying *S*'s goals plausibly can be regarded as *the* attitude that it is rational for *S* to take toward *T.* For, by hypothesis, it is the attitude that it is rational for *S* to take toward *T* when *all* relevant considerations are taken into account (i.e., when *all* of *S*'s goals are taken into account).

Unfortunately for the epistemologist and the philosopher of science, there is a serious drawback to this strategy of allowing any of *S*'s goals, even his nonintellectual goals, to play a role in determining what is the rational attitude for *S* to take toward a theory *T.* The drawback is that it is at least possible for *S*'s nonintellectual goals to play a larger role in determining what attitude it is rational for him to take toward T than either philosophers of science or epistemologists commonly want, given the nature of their projects. Their projects are ones that inquire into what it is rational for us to believe insofar as we are exclusively or at least predominantly intellectual beings, beings whose only concerns or whose dominant concerns revolve around the discovery of truths and the avoidance of falsehoods. But once we allow all of S's goals to be relevant to an assessment of what it is rational for him to believe (in an effort to determine what is the rational attitude for S to take toward a theory T), there no longer is any guarantee that intellectual concerns will prevail. Given S's situation and given S's goals, it *might* turn out that considerations of what will comfort us or considerations of what will make the world a better and safer place or some other such consideration will play a larger role in determining what it is rational for us to believe about T than does our evidence about T. S's situation might be such, for example, that were he to believe what his evidence warrants or were he to use a method of belief-acquisition that is likely over the long run to lead us to easily utilizable truths, the world would be a significantly more dangerous place than would otherwise be the case. And this might make it rational, all things considered, for S *not* to believe what his evidence warrants, or it might make it rational, all things considered, for him not to use a method of belief acquisition that is likely to lead to easily utilizable truths. Even so, if we want an answer to the question of what is *the* rational attitude for S to take toward T, this is a drawback we must accept. It is a drawback that follows straightforwardly from the fact that what we believe affects our practical situation as well as our intellectual situation and that in determining what it is rational, all things considered, for us to believe, the practical consequences of our beliefs are in principle as relevant as the intellectual consequences.[15]

The second qualification has to do with the perspective we are to adopt in trying to get an answer to our question of what is *the* rational attitude for S to take toward a theory T. In particular, we need to make a decision about the perspective from which we are going to evaluate how effectively S's believing T (or disbelieving T or withholding judgment on T) will promote the total constellation of S's goals. However, I already have claimed that there is no principled way to make this decision, since there is no privileged perspective from which to make evaluations about how an individual is pursuing his or her goals. Depending upon our purposes, intentions, etc. it can be appropriate to make these evaluations from the perspective of the individual himself or from the perspective of his community or from some other external perspective, and it can be equally appropriate to give expression to these evaluations by making a claim about the rationality (or irrationality) of S's believing T. So, although we *can* make judgments concerning what is *the* rational attitude for S to take toward a theory T (by taking into account all of S's goals and by making a decision about what perspective to adopt), we should not deceive ourselves into thinking that this is the only perspective that it is appropriate to adopt in making such evaluations.

It is important not to overemphasize the relativism implicit in this conclusion, however. There *is* a kind of relativism implicit in the claim that there is no privileged perspective from which to make evaluations of an individual's beliefs and the intellectual practices, methods, and habits that led to these beliefs, but it is not a vicious, paralyzing kind of relativism, whereby an individual is left without a perspective from which to make decisions about these beliefs, practices, methods and habits. On the contrary, the decision-making process goes on as before. The relativistic conclusion does not affect it at all. Indeed, in one important sense the individual has no choice whatsoever about the perspective from which to make decisions. Inevitably the individual makes decisions from a subjective perspective rather than an objective perspective or an intersubjective perspective. How could it be otherwise?[16] The above conclusion is concerned only with the evaluation of the individual's beliefs, practices, habits, and so on. Moreover, it is only concerned with one kind of evaluation of these beliefs, practices, habits, etc.—that is, evaluations of them as rational or as irrational. Other kinds of evaluations—for example, moral evaluations—may have to be understood in a fundamentally different manner. Thus, the point here is "only" this: Since there is no privileged perspective from which to make evaluations concerning the rationality of an individual's beliefs, we can and do, depending on our purposes and interests, make such evaluations from a variety of perspectives. We can make them from the individual's perspective, but we can make them equally well from a community-based perspective, as well as from a host of other perspectives.

4. A Way to Think about Rational Belief

Where does this leave us? It leaves us, most importantly, with a way to think about questions of rational belief. Part of what has been illustrated by the discussions here—both the discussion of the differences between accounts of scientifically rational beliefs and accounts of epistemically rational beliefs, as well as the discussion of the differences between reliabilist accounts of rational beliefs on the one hand and foundationalist and coherentist accounts of rational belief on the other hand—is that the general conception of rationality I am proposing has considerable heuristic power. It in effect creates a methodology for classifying (and hence comparing, contrasting, and understanding) work in the philosophy of science, work in epistemology, and work in other areas of philosophy that are concerned with rational belief—work that is otherwise hard to compare and contrast. However, my aim here is not simply classificatory. It is also therapeutic, the idea being that philosophical work on rational belief inevitably will suffer until philosophers are sufficiently reflective about what kind of account of rational belief they are trying to develop. Given that claims of rationality are elliptical, it is incumbent upon epistemologists, philosophers of science, and others who are interested in such questions to make clear from the start what notion of rational belief they are trying to explicate. Moreover, the way to make this clear is to specify the point of view from which the evaluations of the individual's beliefs are to be made. The way to specify the relevant point of view, in turn, is by specifying the goal being sought by the individual (for example, whether this be a present-tense intellectual goal or a future-tense intellectual goal or a practical goal or some combination of such goals) and by specifying as well the perspective from which the evaluations of how effectively the individual's beliefs promote this goal are to be made (for example, whether they are to be made from the individual's own perspective or from some external perspective).

Most simply put, my proposal is this. Whenever someone makes a claim about the rationality of an individual's actions or beliefs and, *a fortiori,* whenever one proposes a philosophical account of what is involved in such claims, we can and should ask two questions: Rational for what— i.e., with respect to what goal? And rational from what perspective?[17]

NOTES

1. For a thorough discussion of such proposals, see Alvin Goldman, *Epistemology and Cognition* (Cambridge: Harvard University Press, 1986), especially pp. 103–9.

2. For convenience, the notion of goal here can be construed

broadly as anything that the individual needs or wants. However, this is a matter of convenience only. I make no presupposition about what is an acceptable theory of goals. I make no presupposition, e.g., concerning whether something might not be a goal of an individual even if he neither wants nor needs it. One can "plug" one's favorite theory of goals into the above characterization.

3. I am not assuming here that we have direct control over what we believe, although I do think that we can exercise indirect control over what we believe—say, by improving our intellectual habits, by developing an attitude of critical curiosity, and the like.

4. Thus, the coherentist typically focuses upon what the individual happens to believe and then imagines the individual correcting those beliefs in order to remove any incoherence among the beliefs. The point of removing any incoherence is to create a belief system that, given the perspective of the individual, seems most likely to provide an accurate representation of the world, the presupposition being that an incoherent belief system could not possibly provide an accurate representation of the world. Although it may not be obvious at first glance, traditional foundational accounts also emphasize the perspective of the individual himself (on reflection). Consider Descartes, for example. According to Descartes, it is epistemically rational for an individual to believe only that which is clear and distinct, where in turn God guarantees that what is clear and distinct is true. Moreover, Descartes thinks that if a normal adult is sufficiently careful in his deliberations about various claims, he can distinguish what is clear and distinct from what is not. So, according to Descartes, there is subjectively available to every normal adult a method of belief acquisition—namely, believing only what is clear and distinct—that is guaranteed to provide an accurate (albeit incomplete) representation of the world.

5. For Descartes, the nature of God precludes this being the case. God, being essentially both good and omnipotent, would not permit this. For some contemporary philosophers, it is the nature of belief that precludes this being the case. "What stands in the way of global skepticism of the sense is, in my view, the fact that we must, in the plainest and methodologically most basic cases, take the objects of a belief to be the causes of that belief." Donald Davidson, "A Coherence Theory of Truth and Knowledge," in *The Philosophy of Donald Davidson: Perspectives on Truth and Interpretation*, ed. E. Lepore (London: Basil Blackwell, 1986). See also Davidson, "On the Very Idea of a Conceptual Scheme," *Proceedings and Addresses of the American Philosophical Association* 17 (1973–74), 5–20; Hilary Putnam, "Realism and

Reason," *Proceedings and Addresses of the American Philosophical Association* 50 (1977), 483–98; and Putnam, *Reason, Truth, and History* (Cambridge: Cambridge University Press, 1981), especially chap. 1.

6. However, see Goldman, *Epistemology and Cognition,* especially p. 113.

7. Are there realistic examples in which believing a proposition that on reflection we would take to be false enhances our future prospects of truth? The general problem in finding examples of such propositions is that insofar as we are aware that a proposition has these characteristics, we will not be inclined to believe it (although we perhaps will be inclined to act *as if* we believed it). Even so, here is one possibility. Relative to our present evidence, it *perhaps* is not highly probable that there is a unified set of fundamental laws governing all physical and biological processes, and perhaps we would admit as much if we reflected sufficiently on the state of our present evidence for this claim. Nevertheless, our believing such a claim may increase our prospects for discovering truths; if scientists believe this, for example, they may be encouraged to push on in their research, which in turn is likely to produce more and more insights (even if it is not likely to produce a unified set of fundamental laws).

8. This is not to say that good scientific practice is not also concerned with the present-tense intellectual goal, the goal of now believing truths and now not believing falsehoods. It is only to say that the emphasis is on a method that, it is thought, is likely eventually to lead us to an accurate and comprehensive account of the world. Of course, one way (indeed, ordinarily the best way) in which a method might lead us eventually to an accurate and comprehensive account of the world is by providing us in the interim with an account that is "approximately true." For, presumably, an account of the world that is approximately true will help guide our research in the direction of ever more accurate and ever more comprehensive accounts. For more on the notion of approximate truth, see Ernan McMullin, "The Fertility of Theory and the Unit of Appraisal in Science," in R. S. Cohen, P. K. Feyerabend, and M. W. Wartofsky, eds., *Boston Studies in the Philosophy of Science: Essays in Honor of Imre Lakatos,* vol. 39 (Dordrecht: Reidel, 1976), pp. 681–718.

9. Suppose there is a community in which astrologers are regarded as experts on intellectual matters, much as scientists are so regarded in our community. Would it then follow, given the above characterization of scientific rationality, that the intellectual practices advocated by these astrologers would be *scientifically* rational in this community? No. What this example illustrates is that it is *possible* for there to be a community in which the intellectual practices of "ex-

pert" astrologers are the standard against which other intellectual practices and the beliefs resulting from these practices are often evaluated much as the intellectual practices of expert scientists are so used in our community. Accordingly, it does follow, given the above discussion, that it is *possible* for there to be a community in which it is rational (in a community-based sense) for an individual to believe the results of investigations conducted in accordance with the methods of proof, evidence gathering, etc., recommended by the astrologers of his community, just as it is rational (in a community-based sense) for us to believe the results of investigations conducted in accordance with the methods of proof, evidence gathering, etc., recommended by expert scientists. However, we need not say that these astrological practices are scientifically rational for the people of this hypothetical community. The term 'scientific' can be and should be reserved for the practices endorsed by *our* experts. Thus, the "only" point here (controversial enough in itself) is that what makes scientific methods rational for us—most roughly expressed, the fact that those regarded by us as experts regard these methods as the best methods to employ insofar as we are interested in obtaining over the long run easily utilizable truths—might make astrological methods rational for members of some other community. This is at least possible.

10. It leaves unanswered, for example, the question of what methods of evidence gathering, testing, etc., are in fact used by scientists, and it also leaves unanswered the related question of whether the methods used by scientists might vary from one area of science to another.

11. See, e.g., Laurence BonJour, *The Structure of Empirical Knowledge* (Cambridge, Mass.: Harvard University Press, 1985), who says (p. 183) that "a hypothesis is complex rather than simple in this sense to the extent that it contains elements within it, some of which are unlikely to be true relative to others, thus making the hypothesis as a whole unlikely on a purely *a priori* basis to be true; it is simple to the extent that this is not the case."

12. "Newton's hypothesis was simpler than its predecessors in that it covered in a brief, unified story what had previously been covered only by two unrelated accounts. Similar remarks apply to the kinetic theory of gases." W. V. Quine and J. S. Ullian, *The Web of Belief,* 2nd ed. (New York: Random House, 1978), p. 71.

13. W. H. Newton-Smith, *The Rationality of Science* (Boston: Routledge & Kegan Paul, 1981), p. 227.

14. Relevant here is Ernan McMullin's distinction between *U*-fertility (the as yet untested promise of a theory for opening up new areas of inquiry, meeting anomalies, and so on) and *P*-fertility (the

proven fertility of a theory). The first is a measure of the theory's potential, the second a measure of its performance. McMullin argues that although U-fertility is irrelevant for the epistemic appraisal of a theory, P-fertility is epistemically relevant; the fact that a theory has proven fertile makes it, all else being equal, more likely to be true. This clearly is the way to give fertility its best epistemic run for the money, but it is hard to see why even P-fertility needs to be epistemically relevant. The issue is complicated by the fact that it is not implausible to suppose that P-fertility may very well be a derivative virtue of theories. However, suppose we imagine an example in which this possibility is discounted. Suppose that theory[1] has been the dominant theory for a period of time and that during that time it has proven to be fertile. Now suppose a new theory[2] is developed. Since theory[2] is new, it cannot have P-fertility, but suppose in *all* other respects (including U-fertility) it is the equal of theory.[1] Why, then, should it be the case that the P-fertility of theory[1] necessarily gives us a reason to think that it is more likely to be true than theory[2]? Contrast with Ernan McMullin, "The Fertility of Theory . . . ," especially pp. 686–88.

15. This conclusion is tempered somewhat by the fact that ordinarily (but not necessarily) there are good practical reasons to let intellectual concerns dominate our belief-acquisition practices. For ordinarily (but again, not necessarily), we enhance our chances of satisfying our practical goals by enhancing our chances of having true beliefs—e.g., true beliefs about how to effectively pursue our practical goals.

16. Of course, this is not to say that an individual cannot try to adopt an impartial, objective perspective in making decisions. It is only to say that a decision to adopt such a perspective is itself a decision (if it is a decision at all) that is made from the individual's own perspective, and it is only to say as well that a decision about *how* to be impartial and objective is a decision that is made from the individual's own perspective. It is in this sense that the subjective perspective is inescapable in the decision-making process. Contrast with Thomas Nagel, *The View from Nowhere* (Oxford: Oxford University Press, 1986), especially chaps. 8 and 9.

17. Many of the themes in this essay are developed more extensively in *The Theory of Epistemic Rationality* (Cambridge: Harvard University Press, 1987).

MICHEL FOUCAULT AND
THE HISTORY OF REASON

Gary Gutting

1. Introduction: Foucault's Critique of Reason

In Simone de Beauvoir's autobiography she tells how Sartre first learned, from Raymond Aron, of Husserl's phenomenological method. Aron had been studying Husserl's philosophy in Berlin and was telling Sartre and de Beauvoir about it in a cafe on the rue Montparnasse:

> We ordered the speciality of the house, apricot cocktails. Aron said, pointing to his glass: "You see, my dear fellow, if you were a phenomenologist, you could talk about this cocktail glass and make philosophy out of it." Sartre turned pale with emotion at this. Here was just the thing he had been longing to achieve for years—to describe objects just as he saw and touched them, and extract philosophy from the process.[1]

This incident illustrates a central trait of twentieth-century philosophy: its desire to subordinate the abstract conceptions and arguments of philosophical theories to the richness and surety of a concretely existing world. For Sartre—and the phenomenological and existentialist traditions generally—this concrete world is that given in our immediate experience. For the later Wittgenstein and his followers the focus is on the concreteness of our ordinary linguistic usages. For positivist and post-positivist

philosophers of science in the analytic tradition concreteness is found in the actuality of present or past scientific
practice.

One of the most philosophically exciting things about
the work of Michel Foucault is its development of a new
focus in concrete reality for philosophy. "You see, my dear
fellow," we can imagine someone saying to a 1986 coun-
terpart of the young Sartre, "if you were a Foucaultian,
you could talk about madmen, disease, prisons, and sex
and make philosophy out of it." But the difficulty is in
seeing just how we are to make philosophy out of
Foucault's brilliant historical talk. Although we know that
he was trained as a philosopher and possessed a sort of
dialectical intelligence that might be judged philosophical,
it has been much easier to read him as a historian of
thought or social theorist and critic than as a philosopher
in any fundamental sense.

In fact, however, Foucault's work was profoundly and
essentially philosophical, the product of what he himself
called a "philosophical *ethos*" aimed at achieving a "critical ontology" of human reason. Specifically, Foucault saw
his work as a legitimate twentieth-century successor to
Kant's eighteenth-century critique of reason. In Foucault's
view, the key to Kant's project can be found in his response to the *Berliner Monatschrift's* question, "What is
Enlightenment?" Kant's answer was that enlightenment is
man's release from his "inability to make use of his understanding without direction from another," an inability
that was to be overcome by finding the courage to use our
own reason rather than submitting it to books, pastors,
physicians, and other external authorities. Kant felt that
his own age was the beginning of reason's emergence as
an autonomous force and so required a careful assessment
of its precise scope and limits. As Foucault puts it:

> it is precisely at this moment that the critique is necessary,
> since its role is that of defining the conditions under which
> the use of reason is legitimate in order to determine what

can be known, what must be done, and what may be hoped. Illegitimate uses of reason are what give rise to dogmatism and heteronomy, along with illusion; . . . it is when the legitimate use of reason has been clearly defined in its principles that its autonomy can be assured.[2]

As Foucault sees it, Kant's effort at critique was dominated by the idea—characteristic of the "modern thought" that abruptly emerged at the end of the eighteenth century—that "the limits of knowledge provide a foundation for the possibility of knowing."[3] Thus, on Kant's account, that which limits the scope of reason—our dependence on the forms of sensibility and the categories of the understanding—are inevitable consequences of necessary, *a priori* structures that define the very possibility of knowing. This account depends on a sharp distinction between the *form* of knowledge, which has a transcendental status, free from all contingency and historicity, and the *content* of knowledge, which is the empirical locus of historical contingency. However, it became apparent—especially after Hegel—that even the forms of knowledge (e.g., the fundamental categories) were themselves subject to the vicissitudes of history, so that Kant's sharp distinction of the transcendental and the empirical could not be maintained. As Foucault reads it, the history of modern (i.e., post-Kantian) thought is a series of vain efforts to account for knowledge in terms of Kant's framework. Some (empiricists, positivists) assimilated the transcendental to the empirical, reducing knowledge to a natural phenomenon in the world; others (idealists) assimilated the empirical to the transcendental, making knowledge an unconditioned absolute. But, as Kant himself would have anticipated, the first approach could not avoid skepticism and the second falls into dogmatism. Other modern thinkers (especially, Husserl, the early Heidegger, and Merleau-Ponty) accordingly tried to revive some equivalent of Kant's original picture of the knower as an "empirical/transcendental doublet"—both an object in the world and a subject constituting the world of objects. But all such efforts have failed, revealing the intrinsic incoherence and instability of Kant's "doublet" conception. This failure is what Foucault referred to in his misleadingly grandiloquent phrase, "the death of man."[4]

But, even though Kant's way of conceiving knowledge needs to be surpassed, Foucault thinks his project of a critique of reason remains as a task. That is, Foucault accepts the Enlightenment's central goal of the autonomy of reason and thinks that attaining this autonomy requires reflection on the scope and limits of reason. But, in the wake of the failure of Kant's conception of knowledge, the critique of reason must take a new form. We should abandon the attempt to determine the *a priori*, necessary conditions governing the exercise of reason ("formal structures with universal value"). Rather, we should reflect on what *seem* to be universal *a priori* structures of thought to see the extent to which they have in fact a contingent, historical origin and the extent to which reason may free itself from ("transgress") the constraints of these structures.

> Criticism indeed consists of analyzing and reflecting upon limits. But if the Kantian question was that of knowing what limits knowledge has to renounce transgressing, it seems to me that the critical question today has to be turned back into a positive one: in what is given to us as universal, necessary, obligatory, what place is occupied by whatever is singular, contingent, and the product of arbitrary constraints? The point, in brief, is to transform the critique conducted in the form of necessary limitation into a practical critique that takes the form of a possible transgression.[5]

This explains why, for Foucault, philosophy takes the form of historical reflection and why his detailed studies of madness, prisons, sexuality, etc., have philosophical import. In all his historical studies, Foucault examines major loci of "rationalization"; i.e., instances in which there seem to have been decisive applications of rational methods to our understanding and evaluation of key aspects of human existence. Thus, *Folie et déraison* treats the early nineteenth-century "enlightened" understanding of insanity (and hence of its opposite, sanity or reason); *The Birth of the Clinic* traces the emergence of "scientific" medicine; and *The Order of Things* situates and analyzes the post-Kantian critical approach to knowledge. Similarly, *Discipline and Punish* and the *History of Sexuality* study allegedly enlightened views of crime and sex.

Taken as a whole, Foucault's project of historical reflection represents an important reconception of the cultural role of philosophy. He gives us the traditional philosophical goal of grounding theoretical and practical knowledge in an understanding of the essential, universal structures of thought and reality and instead applies the philosopher's analytic and synthetic skills to the task of uncovering and, when possible, dissolving contingent, historical constraints on thought. He thus abandons the venerable but empty pretensions of philosophy to provide a privileged access to fundamental truths. But, at the same time, he offers a more concrete and effective approach to the equally venerable goal of liberating the human spirit.

In what follows, I plan to offer a preliminary understanding of some central aspects of Foucault's philosophical project, first by situating it in a context—that of recent French history and philosophy of science—necessary for appreciating it and, second, by explicating the "archaeological" method that Foucault employs as the instrument of his philosophical reflection.

2. Foucault and the Bachelard/Canguilhem Network

Foucault himself notes that there has been in France a long and important tradition of philosophical reflection on the history of reason. In this tradition, going back at least to Comte, historical reflection on reason has been carried out through a philosophically informed and attentive history of science. (By contrast, in Germany, from the left Hegelians to the Frankfurt School, such reflection has focused on political and social developments.) Particularly important, though little known outside of France, is a twentieth-century "network" (Foucault's French term is "filiation") of historians and philosophers of science that includes Koyré, Cavaillès, Bachelard, and Canguilhem. On the one hand, Foucault sees this network as a center for specific studies relevant to the Enlightenment's concern about the autonomy of reason:

> Works such as those of Koyré, Bachelard, Cavaillès, or Canguilhem . . . served as important philosophical starting points to the extent that they shed light on different aspects of this question of the Enlightenment, which is essential to contemporary philosophy.[6]

On the other hand, he also sees the network as an alternative to the phenomenological and existential philosophies that dominated French thought (especially as viewed from the outside) after World War II. Thus, Foucault insists on a basic division in recent French thought "that separates a philosophy of experience, of sense, and of subject [from] a philosophy of knowledge, of rationality, of concept,"[7] the first being the philosophy of Sartre and Merleau-Ponty, the second that of Cavaillès, Bachelard, and Canguilhem.

Foucault, a student of Canguilhem, explicitly includes himself in this second group.[8] This is not to say that he did not feel the impact of phenomenology and existentialism. He was also a student of Merleau-Ponty (whose interpretation of Husserl he implicitly accepts in *Les mots et les choses*) and his early (1954) essay on Binswanger shows strong Heideggerian sympathies.[9] But Foucault soon concluded that phenomenology was not adequate to some central philosophical problems. Specifically, it could not offer an adequate account of language, whereas a structuralist approach, which gave no role at all to the constitution of meanings by a transcendental subject, was extremely illuminating. Similarly, a linguistic-structuralist approach to the unconscious was far superior to the phenomenological explications offered by existential psychiatry. So, like many French intellectuals, Foucault was, by the end of the 1950s, quite finished with the phenomenological approach to philosophy. But, unlike many others, he had readily available an alternative philosophical approach: that of Bachelard and Canguilhem. Consequently, to understand the precise form taken by Foucault's philosophical history of reason, we need to understand its relations to the work of these two thinkers.

The conceptual basis of this work lies in Bachelard's philosophy of science, developed in a series of books written from the 1920s through the 1950s. According to Bachelard, reason is best known by reflection on science, and science is best known by reflection on its history. The first thesis derives from Bachelard's conviction that the structures of reason are apparent not in abstract principles but in the concrete employments of reason. Norms of rationality are constituted in the very process of applying our thoughts to particular problems, and science has been the primary locus of success in such applications. The proof of the second thesis—that science is best known through its history—lies in the repeated refutation of *a priori* philosophical ideals of rationality by historical scientific developments. Descartes, for example, held that rational science must be grounded in clear and distinct intuitions of the essential properties of matter. This view is refuted by the fact that matter, as described by twentieth-century physics and chemistry, is simply not available to our intellectual intuition. We know it only through the indirections of hypothetico-deductive inference from data that are themselves mediated by complex instruments.[10] Similarly, Kant's formulation of a transcendental, *a priori* analytic of principles that regulate all employments of reason collapsed with the triumph of theories (relativity and quantum mechanics) based on the denial of such Kantian principles as the permanence of

substance, which required a continuity of energy inconsistent with quantization.[11] What initially seem to be *a priori*
constraints on thought as such turn out to be contingent
conditions derived from philosophers' inability to think
beyond the framework of present science.

There are, then, no viable accounts of rationality except
those derived from the historical developments of scientific reason. To understand reason, philosophy must "go
to the school of science." Here, as elsewhere (e.g., the
development of metaphysical theories), the achievements
of science are the dynamics behind all philosophical understanding. "Science in effect creates philosophy."[12]

The rationality that philosophy tries to discover in the
history of science is no more fixed and monolithic than
that history itself. As we shall see shortly, Bachelard finds
sharp breaks in the history of science and corresponding
changes in the conception of reason. Moreover, Bachelard
reminds us that there is, strictly speaking, no such thing
as the history of science, only various histories of different
regions of scientific work. Correspondingly, philosophy
cannot hope to uncover a single, unified conception of
rationality from its reflection on the history of science;
it will find only various "regions of rationality" ("les
régions rationelles").[13] Thus, Bachelard himself analyzes
(in *Le rationalisme appliqué*) the rationalities implicit in
nineteenth-century theories of electricity and of mechanics. He agrees that the history of science tends to the
integration of diverse regions of rationality but sees no
place for a "science in general" to which would correspond a "general rationality."

Because of his demand that the philosopher of science
work from the historical development of the sciences, the
center of Bachelard's philosophy of science is his model of
scientific change. This model, which also provides Bachelard's account of the nature of scientific progress, is built
around three key epistemological categories: epistemological breaks, epistemological obstacles, and epistemological
acts.

Bachelard employs the concept of an epistemological break in two contexts. First, he uses it to characterize the way in which scientific knowledge splits off from and even contradicts commonsense experiences and beliefs. This sense of "break" is fundamental for Bachelard, since he sees it as constituting science as a distinctive cognitive realm. He sums up his view on this topic in the final chapter of his last book on the philosophy of science: "We believe, then, that scientific progress always reveals a break [*rupture*], constant breaks, between ordinary [*commune*] knowledge and scientific knowledge."[14] Bachelard illustrates this claim with several examples that we can use to elucidate the key features of epistemological breaks. He finds one very simple example in a chemistry text's comment that glass is very similar to wurtzite (zinc sulfite). The comparison is one that would never occur to common sense, since it is not based on any overt resemblance of the two substances but on the fact that they have analogous crystalline structures. Thus, science breaks with ordinary experience by placing the objects of experience under new categories that reveal properties and relations not available to ordinary sense perception.

But we should not think of scientific breaks as merely a matter of discovering new aspects of ordinary objects, of taking up where everyday experience leaves off, as a telescope reveals stars not visible to the naked eye. New scientific concepts are required to give an adequate account of even familiar facts. This is very nicely illustrated by the case of Lamarck's futile efforts to use his exceptional observational abilities to develop an account of combustion in opposition to Lavoisier's. His approach was to note carefully the sequence of color changes a piece of white paper undergoes when burned. On the basis of such observations, Lamarck interpreted combustion as a process whereby the "violence" of the fire "unmasks" the fundamental, underlying color of the paper (black) by stripping away successive chromatic layers. Bachelard argues that Lamarck's idea here is not merely wrong in the ordinary way of an incorrect scientific hypothesis. Rather, it is essentially anachronistic because it is based on immediate phenomenal experiences that Lavoisier had already shown to be inadequate for the task of understanding combustion. "The time for direct, natural observation in the realm of chemistry has passed."[15]

The second sort of epistemological break is that which occurs between two scientific conceptualizations. For Bachelard, the most striking and important such break came with relativity and quantum theory, which he saw as initiating a "new scientific spirit." This "new spirit" involved not only radically new conceptions of nature but also new conceptions of scientific method (e.g., new criteria of explanatory adequacy). Bachelard's detailed treatments of this topic (in, e.g., *La valeur inductive de la relativité* and in *Le nouvel esprit scientifique*) preceded by two or three decades similar discussions by Anglo-American historians of science such as Kuhn and Feyerabend.

The language of epistemological "breaks" suggests that there is something to be broken, a barrier that must be shattered. Bachelard follows out this suggestion with his notion of an *epistemological obstacle*. An epistemological obstacle is any concept or method that prevents an epistemological break. Obstacles are residues from previous ways of thinking that, whatever value they may have had in the past, have begun to block the path of inquiry. Common sense is, of course, a major source of epistemological obstacles. Thus, the animism of primitive common sense, which inclined people to explain the world by analogy with vital processes (sex, digestion, etc.) was an obstacle to the development of a mechanistic physics. Likewise, the still strong commonsense idea that phenomena must be the attributes of an underlying substance blocked the rejection of the ether as the locus of electromagnetic waves. More generally, Bachelard regards the commonsense mind's reliance on images as a breeding ground for epistemological obstacles. Images may have heuristic use in science, but they have no explanatory force; and if they do their job properly, they are eventually eliminated from scientific thought. Thus, of Bohr's planetary model of the atom, Bachelard says: "The diagram of the atom provided by Bohr . . . has . . . acted as a good image: there is nothing left of it."[16] But epistemological obstacles may also arise from successful scientific work that has outlived its value. The most striking such cases occur when the concepts and principles of an established theory lead us to regard new proposals as obviously absurd; e.g., the counterintuitive feel of quantum mechanics' rejection of classical determinism. But previously successful scientific methods can also become epistemological obstacles. For example, the emphasis on direct observation that led in the seventeenth century to major breaks with Aristotelian science became an obstacle to eighteenth century developments of atomic theories. Finally, traditional philosophy, with its tendency to canonize as necessary truths the contingent features of one historical period of thought, is another major source of epistemological obstacles.

The views and attitudes that constitute epistemological obstacles are often not explicitly formulated by those they constrain but rather operate at the level of implicit assumptions or cognitive and perceptual habits. Consequently, Bachelard proposed to develop a set of techniques designed to bring them to our full reflective awareness. He spoke of these techniques as effecting a "psychoanalysis" of reason. Bachelard's use of this term signals his aim of unearthing unconscious or semiconscious structures of thought, but it does not express a commitment to the details of Freudian theory.

The concept of an *epistemological act* counterbalances that of an epistemological obstacle. Whereas epistemological obstacles impede scientific progress through the inertia of old ideas, "the notion of epistemological acts corresponds to the leaps [*saccades*] of scientific genius that introduce unexpected impulses into the course of scientific development.'"[17] An epistemological act is not, however, just a change; it has a positive value that represents an improvement in our scientific accounts. There are, accordingly, different values that must be accorded to different episodes in the history of science. Consequently, Bachelard holds that writing history of science is different from writing political or social history. In the latter case, "the ideal is, rightly, an *objective* narration of the facts. This ideal requires that the historian *not judge;* and, if the historian imparts the values of his own time in order to assess the values of a past time, then we are right to accuse him of accepting 'the myth of progress.' "[18] But in the case of the history of the natural sciences, progress is no myth. Present science represents an unquestionable advance over its past, and it is entirely appropriate for the historian of science to use the standards and values of the present to judge the past. Application of these standards results in a sharp division of the scientific past into *"l'histoire perimée"* (the history of "outdated" science) and *"l'histoire sanctionée"* (the history of science judged valid by current standards).

Georges Canguilhem points out[19] that this Bachelardian writing of the history of the past on the basis of the present is not equivalent to the now generally disdained "Whiggish" approach to the history of science. For one thing, Bachelardian history does not try to understand past science in terms of present concepts. It realizes the need to explicate the past in its own terms. For another, there is no assumption of the immutable adequacy of present science. Precisely because they are scientific, the present achievements by which we evaluate the past may themselves be surpassed or corrected by future scientific development. Our evaluation of the past in terms of the present is, in Canguilhem's words, not the application of "a universal touchstone" but "a selective projection of light on the past."[20]

But, even though all scientific results are open to revision and some can be definitively rejected, others must be accepted as permanently valid achievements. Thus, Bachelard says that phlogiston theory is "outdated [*perimée*] because it rests on a fundamental error." Historians who deal with it are working "in the paleontology of a vanished scientific spirit."[21] By contrast, Black's work on caloric, even though most of it has long been jettisoned, did yield the permanent achievement of the concept of specific heat. "The notion of *specific heat*—we can assert with equanimity—is a notion that is *forever* a scientific notion. . . . One may smile at the dogmatism of a rationalist philosopher who writes 'forever' regarding a scholastic truth. But there are concepts so indispensable in a scientific culture that we cannot conceive of being led to abandon them."[22]

How is this idea of unalterable progress consistent with Bachelard's insistence that all scientific results are open to revision? How can an achievement be "permanent" and at the same time open to correction in the wake of an epistemological break? Bachelard's response is that an epistemological break is not merely the rejection of past science but also a preservation, via reformulation, of old ideas in a new and broader context of thought. Specifically, past results are replaced by generalizations that reject them as unconditionally correct but preserve them as correct under certain restricted conditions. Bachelard finds a model here in the development of non-Euclidean geometry. This development refutes the claim that the Euclidean postulates express the sole truth about geometry but at the same time presents these postulates as defining one exemplification of a more general class of geometries (i.e., Euclidean geometry is the particular geometry possessed by a space of zero curvature). In just the same way, "Newton's astronomy can . . . be seen to be a special case of Einstein's 'pan-astronomy.'"[23] This is so not merely because, to a certain approximation, Newtonian calculations yield the same numbers as Einsteinian calculations but because key Newtonian concepts such as mass and velocity can be shown to be special simple cases of the corresponding Einsteinian concepts. Bachelard characterizes this process of replacement by generalization as "dialectical," not in the Hegelian sense of a synthesis of opposites but in the sense of a process of conceptual expansion whereby what previously appeared to be contraries (e.g., Euclidean and Lobachevskian geometries) are seen as complementary possibilities. Earlier concepts are not mysteriously "sublated" into a higher unity but are *rectified* (corrected) on the basis of superior successor concepts that allow us to explain precisely the extent to which they are applicable.

This account of scientific change allows Bachelard to reject the *continuity* of science and still accept its *progress*. Science develops by a series of epistemological breaks that make it impossible to regard its history as a linear accumulation of truths within a single conceptual framework. The conceptual framework of science at one stage will be rejected as erroneous at later stages. Nonetheless, some of its results may be permanent scientific achievements in the sense that they will be preserved as special cases within all subsequent scientific frameworks. Each successive framework will represent progress over its predecessors in the sense that it has a more general perspective from which the range of validity of previous perspectives can be assessed.

One simple but informative way of looking at Foucault as a philosopher is to regard him as a Bachelardian who extended his master's project for a philosophically informed and attentive history of science into new areas. Bachelard's own work was limited to the history of physics and chemistry. Canguilhem developed a Bachelardian approach to the biological and medical sciences. Foucault moved on even further to take account of the social sciences. This simple schema does reflect a rough direction of development from Bachelard through Foucault. It also explains how Foucault was able to develop his own original work without taking explicit issue with the work of his mentors. The schema also fits in well with Foucault's own comments at the end of *L'archéologie du savoir* on the relation of his work to that of Bachelard and Canguilhem. The latter, he says, dealt with disciplines at the "threshold of scientificity" and so had the goal of discovering "how. . . a science was established over and against a prescientific level, which both paved the way and resisted it in advance, how it succeeded in overcoming the obstacles and limitations that still stood in its way."[24] Such an enterprise Foucault characterizes as "epistemological history of science." His own project—"archaeological history of science"—is concerned with disciplines at an earlier stage, "the threshold of epistemologization," and hence focuses not on the specific methodological procedures and norms that constitute a discipline as a science but with the more general standards (e.g., of coherence and verification) that characterize even nonscientific bodies of knowledge.

Also in favor of the simple interpretation of Foucault as a Bachelardian is the close correspondence of some of Foucault's most central positions to Bachelard's views. Thus, Foucault's critical project of showing the contingent nature of what present themselves as necessary *a priori* limits on knowledge reflects Bachelard's insistence that philosophical *a prioris* derive merely from philosophers' inability to think beyond the categories of current science. Similarly, Foucault's famous championing of discontinuity has very close affinities to Bachelard's notion of epistemological break, and his strong distrust of global accounts and emphasis on local differences and anomalies echo Bachelard's "regions of rationality." Even Foucault's basic project of an archaeological uncovering of the "deep structures" of knowledge seems closely tied to Bachelard's idea of a "psychoanalysis" of knowledge.

Nevertheless, the simple picture of Foucault as a

Bachelardian is ultimately simplistic and untenable. It ignores the fundamental transformation Foucault effected in Bachelard's project of a history of reason. This transformation has two primary aspects, both closely tied to Foucault's reconstrual of the meaning of a critique of reason.

First, Foucault's suspicion of reason goes much deeper than Bachelard's. Bachelard did recognize obstacles and errors as an intrinsic part of reason's history. Hence, for him, as Foucault notes, "what we are to examine essentially is a reason whose autonomy of structures carries with itself the history of dogmatisms and despotisms—a reason which, consequently, has the effect of emancipation only on the condition that it succeeds in freeing itself [from] itself."[25] But for Bachelard the process of reason's self-emancipation is a linearly progressive one—as perhaps it is if we look solely at the physical sciences. But once, like Foucault, we take medical and social-scientific applications of reason as objects of our historical reflection, it becomes much more difficult to sustain the progressivist picture. For example, Foucault's close study (in *Folie et déraison*) of the development of "scientific" psychiatry in the nineteenth century debunked the standard liberal interpretation of the "unchaining" of the insane. And his *Surveiller et punir* yielded similar results for the nineteenth century prison reform movement. The same sort of objectivity and rationality that free us from the domination of intellectual illusions in our study of nature can, when extended to the study of the self and society, be themselves sources of dominating psychological and social illusions.

Second, and even more fundamentally, Foucault goes far beyond Bachelard in eliminating the Kantian transcendental subject (and all its equivalents and successors) as the protagonist of the history of reason. Admittedly, Bachelard takes some important steps toward this elimination and hence toward getting beyond the framework of the transcendental/empirical doublet. Thus, Bachelard's emphasis on epistemological breaks and on the essential role of error in the development of science undermines the *a priori* necessity of Kant's transcendental categories. Further, Bachelard viewed scientific reality as constituted by concepts and theories technologically embodied in scientific instruments, thus giving a material, antitranscendental reading of "conditions for the possibility of knowledge." But Bachelard still retained the picture of knowledge as rooted in the experience of a subject, even if this subject is no longer an autonomous, individual transcendental ego but the historically conditioned, technologically bound scientific community. Bachelard's conception succeeds only in that it avoids the dogmatism of an absolute idealism. It still falls to the dilemma of either reducing to skeptical naturalism or else returning to the incoherent construal of the subject as somehow both one of the objects in the world and the source of their constitution as objects. Foucault's sharp break with Bachelard on this issue is signaled by his switch from psychoanalysis to archaeology as the central methodological metaphor.

Foucault, accordingly, must be understood as radicalizing Bachelardian history of reason in two dimensions. First, he wants to take full account of reason's intimate ties with the forces of domination (power); second, he wants to eliminate entirely the role of a constituting subject in the history of reason. The former dimension corresponds to Foucault's much discussed significance as a social theorist and critic and is obviously central to his task of philosophically liberating human thought. But the latter dimension is also central to this task. This is so first because so many construals of the constituting subject (from Kant's transcendental ego to Hegel's absolute to Comte's progressively evolving human community) impose totalizing and hence totalitarian unities of meaning on human history. Further, viewing the history of reason as centered on the activities of an individual or social subject prevents us from recognizing another and more fundamental level of the history of reason—the enunciative or archaeological level—where there operate constraints that entirely escape subject-centered historical analysis. We now turn to this archaeological level of analysis.

3. Foucault's Archaeology of Thought

With one exception, all of Foucault's major books are histories of aspects of Western thought and culture; and the exception, *The Archaeology of Knowledge (AK)*, is a methodological reflection on this historical work. It would be too much to claim that Foucault's histories were, from the beginning, part of a single coherent project. Foucault himself admits that it was only retrospectively (from the standpoint of *AK*) that he saw fully the unity of his earlier historical case studies; and, he admits, this unity appears only through some reconstruction and reinterpretation of his earlier efforts. It is, further, clear that the historical works after *AK* involve new emphases, particularly regarding the relation of knowledge and power. Nonetheless, Foucault's central idea—of a history of thought independent of the mental lives of human subjects—is crucial for all his historical works; and this idea is most clearly and fully developed in the methodological discussions of *AK*. I will, accordingly, center my treatment of Foucault on the history of thought on these discussions.

The subject of history has, since its earliest beginnings, been the human subject. The first inscriptions we regard as "historical writings" tell deeds of royal persons; and, until recently, historians have primarily devoted themselves to describing and explaining the beliefs, intentions, and actions of human subjects. History has been the history of human thoughts and actions. Foucault notes that there has been a recent and fruitful trend toward a new sort of history (particularly associated with the French *An-*

nales school), history written not in terms of deeds of men but of factors such as geography, climate, and ecological systems that provide the historical "space" in which human agents act. Foucault's goal is to write this sort of nonsubject-centered history for human thought.

Like any historical inquiry, Foucault's begins with *documents,* collections of statements that we have received from our ancestors. Ordinary history—and especially the history of ideas—sees documents as clues to the intentional acts (beliefs, thoughts, desires, feelings) of those who produced them. It uses the objective linguistic data of statements to reconstruct the inner life of subjects. Foucault, by contrast, proposes to take statements as objects of study in their own right, making no effort to use them as means to revive the thoughts of the dead. This is why he calls his enterprise an "archaeology"—an objective analysis of linguistic "artifacts." We are, of course, already familiar with two (nonhistorical) areas of inquiry that treat statements in their own right: grammar, which defines the conditions under which a statement is meaningful, and logic, which specifies what can and cannot be consistently added to a given set of statements. But it is obvious that the set of statements actually made in a given domain and epoch is a very small subset of those permitted by grammar and logic. Ordinarily, we explain the vast number of grammatically and logically possible statements that are *not* made on the basis of the experiences, beliefs, and intentions of subjects. We do not speak of Jupiter hurling thunderbolts because we do not believe in him; the ancient Greeks did not speak of space travel because they had no experience of it; the Victorians suppressed certain aspects of sexuality out of shame. Foucault suggests that in many fundamental cases the explanation for such linguistic gaps is rather that statements are subject to a further set of rules (neither grammatical nor logical) to which speakers unwittingly conform. A set of statements governed by such a set of rules constitutes what he calls a *discursive formation.*

More fully, Foucault regards a discursive formation as involving four basic elements: the *objects* its statements are about, the kinds of cognitive status they have (what Foucault calls their *enunciative modality*), the *concepts* in terms of which they are formulated, and the *themes* (theoretical viewpoints) they develop. However, we should not think of a given discursive formation as defined by a unique system of objects, a single enunciative modality, a distinctive conceptual framework, or a consistent set of themes or theories. The same discursive formation will be a vehicle for discourse about different systems of objects, categorized in terms of different conceptual frameworks; and its statements will have a variety of enunciative modalities and may develop very diverse theoretical viewpoints.

Accordingly, a discursive formation is not distinguished by any unity (of objects, concepts, method, etc.) provided by its elements. Rather, a discursive formation is a "system of dispersion" for its elements: it defines a field within which a variety of different, even conflicting, sets of elements can be deployed. Thus, the unity of a discursive formation is due entirely to the *rules* that govern the formulation of statements about different systems of objects, exhibiting different sorts of cognitive status, employing different conceptual frameworks, and expressing different theoretical viewpoints.

Foucault's general analysis of discursive formations consists of a detailed classification of these rules. They fall into four types, each corresponding to one of the elements of a discursive formation. Thus, there are rules for the formation of objects, for the formation of enunciative modalities, for the formation of concepts, and for the formation of theoretical strategies. Foucault's complex typology of these rules provides an invaluable methodological guide to his historical case studies in *Folie et déraison, Surveiller et punir,* and *Les mots et les choses.*

To understand the import of Foucault's project for an archaeology of knowledge, we need to get clear on two key points: the precise level on which archaeological analysis (the analysis of discursive formations) take place and the nature of the statements that comprise discursive formations. First, let us look at the level of archaeological analysis. On the one hand, archaeology is not concerned with textual analysis, with specific questions about what particular words mean or how particular statements are logically or rhetorically connected. Archaeological analysis "remains anterior to this manifest level" of specific linguistic usage and does not account for the specific details of a particular text; "it leaves the final placing of the text in dotted outline."[26] Foucault also makes this point by distinguishing the *discursive relations* with which archaeology is concerned from the *secondary relations* (grammatical, logical, rhetorical) that govern the concrete uses of languages. But he is even more insistent that archaeological analysis is not an access to a nondiscursive reality that lies outside of and grounds the discursive formation. It does not operate at the level of the *primary relations* that, "independently of all discourse or all objects of discourse, may be described between institutions, techniques, social forms, etc." (*AK,* 46). Put another way, if archaeology does not deal with concrete *words,* neither does it deal with *things themselves* (*AK,* 48). It remains within discourse, but at its borders (which it is concerned to define as clearly and precisely as possible) rather than at its interior. Foucault also makes it clear that his commitment to an archaeological approach does not mean that he rejects the alternative approaches (which we might label "linguistic" and "ontological"). He allows that there is place for the linguistic analysis of a term (of, e.g., what 'melancholia' meant in the seventeenth century) and even that one might write a "history of the referent" that would aim to "uncover and free . . . prediscursive experiences from the tyranny of the text" (*AK,* 47). But whatever the value of these sorts of enterprises, Foucault's archaeology deals neither with prediscursive experience of things nor with the verbal forms produced by discourse. It focuses on "a group of rules . . . [that] define not the dumb existence of a reality nor the canonical use of a vocabulary, but the ordering of objects" (*AK,* 49). Without denying that discourse is composed of signs or that signs can be used to designate things, he insists that there is more to discourse than this. "It is this 'more' that we must reveal and describe" (*AK,* 49).

We next turn to the nature of the statements that comprise discursive formations. Since archaeological analysis is distinct from both grammar and logic, Foucault refuses to identify the statements with which he deals with either sentences—i.e., the units of grammatical analysis—or propositions—i.e., the units of logical analysis. Indeed, he finally concludes that a statement is not really any kind of linguistic *unit* at all but is rather a *function.* To understand what Foucault has in mind here, we need to get some perspective on his way of thinking about language. A language, he says, echoing Saussure, is "a collection of signs defined by their contrasting characteristics and their rules of use" (*AK,* 85). From such a collection we can, of course, form numerous particular series of signs. A given series will be a sentence or a proposition, depending on whether it conforms to the grammatical and logical rules that govern these grammatical units. Further, whether a series of signs is a sentence or a proposition is *entirely* determined by reference to the relevant sets of rules; it does not require that the series have any relation to other series of signs (on the same linguistic level). There could in principle be a language with only one sentence or one proposition. By contrast, a series of signs is a *statement* only if it is related to other series of signs, which series constitute the statement's *associated field* (cf. *AK,* 98). Indeed, the fact that it is a statement and the precise statement that it is are entirely determined by the complex set of rules whereby it is related to other series of signs (which, by virtue of this same set of relations, are themselves statements). Thus, a statement is not a linguistic unit, as are sentences and propositions, since it has no reality as a statement prior to its inclusion within a rule-governed system. Thus, like a mathematical function, a statement is entirely defined by the relations between a set of elements. The other meaning of 'function' is also relevant: a series of signs is a statement precisely because it has a place—a role or a function—within a system.

Linguistic units such as sentences and propositions (and other entities such as graphs, diagrams, and formulas) will also typically be statements; not, of course, simply because they are sentences, propositions, etc., but because they are part of sign systems. As such, they will be open to analysis not only at the standard levels of logic and grammar but also at the *enunciative* level—in terms of their enunciative function ('enunciative' since 'statement' translates the French 'énoncé'). The enunciative level is not unconnected to other semantical levels. Indeed, in important respects it is presupposed by them. Consider, for example, the case in which we say that a proposition—e.g., 'The present king of France is bald'—is false because it has no referent. This will be so only if 'The present king of France is bald' is a statement that belongs to a factual historical discursive formation. If it instead belongs to a fictional discursive formation (say of a novel written about the days of Charles the Bald), then the corresponding proposition will have a referent (in the fictional domain) and indeed be true. Similarly, when we say that a certain string of words (say, 'a way a lone a last a loved a long the') is meaningless, this is so only if we are assuming that the string occurs in some ordinary context and is not part of an experimental literary work. Thus, a group of signs' status as a statement is relevant to whether or not it is (as a sentence or a proposition) true or even meaningful. Specifically, questions of truth and meaning depend on the nature of the relevant discursive formation's domain of objects and relations (which Foucault calls its *referential*).

The nature of Foucault's statements can be clarified by comparing them to the speech-acts that are the concern of language analysts such as Austin and Searle. Although Foucault initially maintained (in *AK*) that statements cannot be identified with speech-acts, he later admitted, in correspondence with Searle, that "I was wrong in saying that statements were not speech acts."[27] But exactly what are we to make of this admission, which Foucault does not discuss in any published text? Statements are speech-acts in the sense that they are "things done with words" or moves in a language game. That is, the analytic category of speech-act is extensionally equivalent to that of the statement. Both are general linguistic categories that include not only expressions of sentences and propositions but also of other linguistic units such as graphs, tables, and gestures. But, as Foucault also suggests in his correspondence with Searle, his concern with statements is very different from the concern a language analyst has with speech-acts. In my view this difference can be best put this way.[28] The analyst works at the level of *meaning;* i.e., the level of the implicit understanding of a language possessed by those who use it. (Hence the standard method of asking, "What do we say if . . . ?" or "What would we say if . . . ?" Such analysis is concerned with distinguishing and describing the functions that speech-acts have within a language. But Foucault, as an "archaeologist," is rather concerned with a structure of relations between statements that is not available to speakers' reflections on the meaning of what they say. He wants to look at statements from the outside and describe the relations that define the field in which various sorts of statements are able to perform their linguistic functions and hence have various meanings.

The rules and relations that govern statements have, as we would expect, nothing to do with the beliefs or intentions of human subjects:

> The analysis of statements operates . . . without reference to a *cogito*. It does not pose the question of the speaking subject. . . . it is situated at the level of the 'it is said'—and we must not understand by this a sort of communal opinion, a collective representation that is imposed on every individual . . . but we must understand by it the totality of things said, the relations, the regularities, and the transformations that may be observed in them. . . . (*AK,* 122)

Of course, all statements are made by individual speakers, but in making a statement a speaker takes up a position that has already been defined—quite apart from his mental activity—by the rules of the relevant discursive formation. Foucault does hold that every statement has a *subject* (not in the grammatical sense but in the sense of a discursive source). But this subject is not any "speaking consciousness" (which will at most be the author of a particular formulation of the statement) but rather "a position that may be filled in certain conditions by various individuals" (*AK,* 115). This position is, of course, established by the rules of the discursive formation.

4. Foucault's Distinctiveness

The distinctiveness and power of Foucault's archaeological approach is perhaps most apparent in four key areas where it differs from traditional history of ideas: (1) its indifference to questions of originality; (2) its view of the relation of thought and discourse to political, social, and economic events and institutions; (3) its attitude toward discontinuities in the history of thought; (4) its way of situating science in the space of intellectual history.

(1) History of ideas is dominated by the two poles of
the old and the new: "in every *oeuvre,* in every book, in the
smallest text, the problem is to rediscover the point of
rupture, to establish, with the greatest possible precision,
the division between the implicit density of the already-
said, a perhaps involuntary fidelity to acquired opinion,
the law of discursive fatalities, and the vivacity of creation, the leap into irreducible difference" (*AK,* 142). As a
result, such history is concerned with ordering the
thoughts of individuals in a single great chronological series, with each member of the series characterized by its
degree of resemblance to previous members of the series.
Thus, a primary concern is finding out who was the true
originator of a given thought and who merely repeated or
creatively modified it; or, as Foucault puts it, "determining those degrees of nobility that are measured here by the
absence of ancestors" (*AK,* 143). Although, following
Bachelard and Canguilhem, Foucault is disdainful of such
searching for intellectual precursors, he does allow for its
significance within certain precisely defined, sufficiently
homogeneous fields of discourse. However, he insists that
for the archaeology of knowledge "the originality/banality
opposition . . . is not relevant" (*AK,* 144). The archaeologist is concerned only with what Foucault calls the "regularities" of discursive practices (enunciative regularities).
These are the patterns defined by the relation of any given
statement to other statements. These patterns (expressed
in the rules of the discursive formation) define the field in
which all statements, from the most creative to the most
banal, emerge. The archaeologist is concerned only with
what a statement can tell him about the rules of the discursive formation he is studying. Since the most original
statement embodies the relevant rules no more and no less
than its hackneyed repetitions, the question of innovation
is of no concern to him.

(2) Archaeology provides a distinctive approach to the
relations between discourses and nondiscursive domains

such as "institutions, political events, economic practices and processes" (*AK,* 162). History of ideas explains such relations via either symbolic or causal analysis. The former sees a discourse (e.g., that of clinical medicine in the late eighteenth century) and nondiscursive factors (e.g., the political, economic, and institutional developments of the eighteenth century) as sharing a common form or meaning in virtue of which each reflects the other. The latter tries to "discover to what extent political changes, or economic processes, could determine the consciousness of scientists" (*AK,* 163); e.g., how nineteenth-century industrial capitalism's need for large numbers of workers caused the medical profession to think and speak of the origin and cure of diseases in social terms. Archaeology is concerned rather with the discursive formation as a condition of the possibility of such symbolic and causal connections. Specifically, with respect to the latter, this means that archaeology "wishes to show not how political practice has determined the meaning and form of . . . discourse, but how and in what form it takes part in its conditions of emergence, insertion, and functioning." Thus, an archaeology of early nineteenth-century medicine (carried out in Foucault's *Birth of the Clinic* shows how social phenomena such as conscripted armies and public-assistance hospitals for the poor provided the context for the emergence of the statistical norms of health discussed by clinical medicine. On another level, the special authority of the doctor (as "virtually the exclusive . . . enunciator of [medical] discourse"—*AK,* 164) is connected with the nineteenth-century institutions of hospitalization and private practice.

Foucault's archaeological approach to the relation of scientific discourse to nondiscursive factors differs in two very important ways from that of most contemporary sociology of science (e.g., the Edinburgh "strong program"). For one thing, as we have already noted, Foucault is not concerned with the influence of social factors on the *content* of scientific theories (e.g., the influence of seventeenth-century political ideology on Newton's laws of mechanics); archaeology works on the more fundamental level of the definition of basic objects and concepts, the cognitive authority of the scientist, and the social function of science. More importantly, Foucault is reluctant to characterize the relation of society to science in causal terms. Rather, he speaks of social factors as "open[ing] up new fields for the mapping of [scientific] objects" (*AK*, 163) and of a scientific practice as "articulated on [social] practices that are external to it" (*AK*, 164). In this way he seems to be trying to make intellectual room for a discussion of science and society that will connect the two on fundamental levels but not require the reductionists' presupposition of social determinism.

(3) Foucault (and, even more, some of his commentators) has emphasized the central role that discontinuity plays in his approach to the history of thought. At the beginning of *AK*, he says:

> One of the most essential features of the new history is probably this displacement of the discontinuous: its transference from the obstacle to the work itself; its integration into the discourse of the historian, where it no longer plays the role of an external condition that must be reduced, but that of a working concept. . . . It is no longer the negative of the historical reading . . . but the positive element that determines its object and validates its analysis. (*AK*, 9)

This insistence on discontinuity has led many of Foucault's readers to think that his archaeological approach allows no place for gradual transformations or continuous developments, that it sees the history of thought as a series of quantum leaps from one self-contained discursive formation to another. (Compare the similar reaction of readers of Thomas Kuhn to his notion of the incommensurability of paradigms.) But such an idea is clearly a misrepresentation, as we can appreciate by seeing the precise role discontinuity plays in Foucault's conception of history. First of all, the emphasis on discontinuity has special importance as a current strategy for writing the history of thought. Traditional history of ideas has emphasized the continuity of human thought through the centuries by reading it as "homogeneous manifestations of a single mind or of a collective mentality." Undermining this sort of continuity has been a necessary part of the new history of thought that questions the privileged role of the human subject. Accordingly, "beneath the great continuities of thought . . . , one is now trying to detect the incidence of interruptions" (*AK*, 4). For such an enterprise, "the great problem is not how continuities are established, . . . how for so many different, successive minds there is a single horizon. . . . the problem is no longer one of tradition, of tracing a line, but one of division, of limits." (*AK*, 5). But such an emphasis on discontinuity is merely a strategy presently appropriate for the history of thought. Other sorts of history have, on the contrary, eliminated the central role of the subject by emphasizing long-term continuities that are independent of the flux of human action. "For many years now historians have preferred to turn their attention to long periods . . . , the movements of accumulation and slow saturation, the great, silent, motionless basis that traditional history has covered with a thick layer of events" (*AK*, 3). Thus, in principle, nonsubject-centered history can (depending on the strategic situation) emphasize either continuity or discontinuity.

Foucault further makes it clear that an archaeology of thought is concerned with changes from one discursive formation to another and that these changes may occur against a background of significant continuities. He agrees that in certain senses archaeological analysis works synchronically: rules of discursive formations may remain the same for long periods and the order the archaeologist discovers in a set of statements may not correspond to the order in which the statements appeared temporally. But, Foucault insists, though "there is a suspension of temporal succession . . . , this suspension is intended precisely to reveal the relations that characterize the temporality of discursive formations" (*AK*, 167). Further, these relations will typically involve the continuity of various discursive elements through a given change. Thus, monetary circulation was an object of both classical analysis of wealth and the nineteenth-century economics that replaces it; and the concept of *character* appears in both classical natural history and in nineteenth-century biology.

So archaeology does not differ from traditional history of ideas by ignoring change and continuity. But it does differ by taking difference and discontinuity as seriously as it does continuity. According to Foucault, traditional history of ideas tries to reduce all apparent discontinuity to a series of incremental changes, all contributing toward a finally achieved enlightenment. Here he seems to have specifically in mind the long-dominant tradition of "Whiggish" history, which presents a cumulative progression of achievements, with the numerous errors and misdirections that undeniably occurred, as unimportant background noise. Foucault rejects this project of "total history," which assumes that the phenomena it deals with are unified around a single center (the progress of mankind, scientific truth) in favor of what he calls "general history" (*AK*, 9–10). The latter allows that its phenomena may form disparate series that cannot be reduced to a unity but without insisting that these series are entirely independent. General intellectual history (perhaps best exemplified by Foucault's *The Order of Things*) seeks to describe the complex interrelations of mutually irreducible discursive formations.

(4) Foucault's account of the relation of discursive formations to the sciences is based on the special sense he gives to the distinction between *connaissance* and *savoir*. By *connaissance* he means (in accord with ordinary French us-

age) any particular body of knowledge such as nuclear physics, evolutionary biology, or Freudian psychoanalysis; thus, *connaissance* is what is found in scientific (or would-be scientific) disciplines. *Savoir*, on the other hand, refers to the discursive conditions that are necessary for the development of *connaissance*, to, in Foucault's words, "the conditions that are necessary in a particular period for this or that type of object to be given to *connaissance* and for this or that enunciation to be formulated" (cf. *AK*, 15, translator's note 2). On Foucault's view, a particular science is the locus of *connaissance*, whereas a discursive formation is the locus of *savoir*. As such, the *savoir* of a discursive formation provides the objects, types of cognitive authority (enunciative modes), concepts, and themes (theoretical strategies) that are necessary for a body of scientific *connaissance*. Or, we might say, a discursive formation provides the preknowledge (*savoir*) necessary for the knowledge (*connaissance*) achieved by a science. This latter locution is justified by Foucault's talk of *savoir* as the "basis" or "precondition" of *connaissance*. But it is important that we not think of *savoir* in this role as an epistemological given (*donnée*), "a lived experience, still implicated in the imagination or in perception" (*AK*, 182). Foucault's *savoir/connaissance* distinction is not a version of the phenomenologist's idea that we begin with uncritical, "immediate knowledge" that is transformed by rigorous method into apodictic scientific knowledge. The *savoir* presupposed by a science is not "that which must have been lived, or must be lived, if the intention of ideality proper to [the science] is to be established." *Savoir* is rather "that which must have been said—or must be said—if a discourse is to exist that complies . . . with the experimental or formal criteria of scientificity" (*AK*, 182). *Connaissance* is an achievement of an individual or a group consciousness and so is naturally the focus of a subject-centered enterprise such as traditional history of science. *Savoir*, by contrast, is the concern of Foucault's archaeology: "Instead of exploring the consciousness/knowledge *(connaissance)*/science axis (which cannot escape subjectivity), archaeology explores the discursive practice/ knowledge *(savoir)*/science axis" (*AK*, 183).

From an archaeological view, a science is just one, localized formation in the "epistemological site" that is a discursive formation. Science neither supersedes nor exhausts the discursive formation that is its background. Thus, an archaeological approach to science differs essentially from standard history of science. It does not, like the traditional approach, proceed on the basis of the assessment of the significance and validity of the past provided by the norms of current scientific practice. Rather, it seeks the origin of epistemic and scientific norms in the relevant discursive formations, seeing such norms not as unquestionable givens for historical reflection but as themselves the outcomes of historical processes (cf. *AK*, 190–91). Here we find another difference that sets Foucault's work off from that of Bachelard and Canguilhem. He proposes a method of writing the history of scientific reason without presupposing present norms of scientific rationality. Because of this, Foucault thinks his archaeological history can provide the basis for a historical critique of scientific rationality, something that is beyond the capacity of ordinary history of science.

Foucault also sees archaeological history as offering a particularly valuable way of approaching the relation between science and ideology. On the one hand, archaeology enables us to see ideology as a natural accompaniment of any science. A science's origin from a discursive formation that also provides intellectual space for nonscientific (e.g., political, religious) practices makes it inevitable that there will be deep similarities and important practical connections between the sciences and the political, economic, religious, etc., ideologies of an era. Thus, to cite a standard example, it should come as no surprise that "political economy has a role in capitalist society, that it serves the interests of the bourgeois class, that it was made by and for that class, and that it bears the mark of its origins even in its concepts and logical architecture" (*AK*, 185). But, on the other hand, archaeology enables us to see that even strong ideological connections need not exclude the scientificity of a discipline: "Ideology is not exclusive of scientificity. Few discourses have given so much place to ideology as clinical discourse or that of political economy: this is not a sufficiently good reason to treat the totality of their statements as being undermined by error, contradiction, and a lack of objectivity" (*AK*, 186). A given scientific discipline will have ideological significance and functions precisely because of the way it is related to other discourses rooted in its discursive formation. But this need not alter the fact that the discipline in itself is governed by norms of scientific objectivity. (Note how Foucault consistently rejects a *teleological* interpretation of the role of ideology in science. It is not a question of someone or some class *using* the science for its purposes but of a common presubjective origin for both the science and the ideology.) It may even be that the ideological function of a science requires that it meet certain standards of objectivity. For example, history designed to further the claims of a religious institution competing for allegiance in a pluralistic society may need to be objective to attain the ideological goal of gaining the respect and attention of nonbelievers. Of course, ideology may also cause defects of objectivity in a science. But, Foucault holds, eliminating the defects need not destroy the ideological connections: "The role of ideology does not diminish as rigour increases and error is dissipated" (*AK*, 186). Thus, archaeology leads away from the standard view that there is a sharp separation between valid science and ideologically influenced inquiry and leads us to see scientific objectivity and ideological bias as two intertwined aspects of a discipline's rootedness in a discursive formation.

In conclusion: Our survey of Foucault's archaeological enterprise shows how it is well suited as the instrument for his project of a critique of reason. First, it uncovers a fundamental level of knowledge that is not related to any constituting subject and so avoids the incoherence of Kant's transcendental/empirical doublet. At the same time, it allows us to take account of the role of contingent social, economic, and political influences on thought, as well as the role of ideology, without requiring a skeptical abandonment of the objectivity of knowledge. Further, Foucault's archaeological method reveals the contingent, historical nature of factors that might, from the viewpoint of the subject, appear to be universal, necessary conditions on knowledge as such; and, in particular, it provides a basis for a critique of scientific norms of rationality. Finally, archaeology's capacity for appreciating radical discontinuities in history enables it to present us with deeply different ways of thinking that can serve as bases for creative "transgressions" of what might otherwise seem to be inviolable limits to our thinking. Thus, Foucault's archaeology constitutes the method of a new, historical form of critical philosophy.

This method, as I noted at the outset, is particularly significant for those of us who see scant prospects for a fulfillment of philosophy's traditional goal of legitimating knowledge-claims and actions via a body of fundamental, essential truths. While eschewing this goal, Foucault still is able to assign philosophy an important role in liberating humankind from arbitrary constraints. To our skeptical age, he offers the hope that, even without the "Truth," we may still be made free.

NOTES

1. Simone de Beauvoir, *The Prime of Life*, trans. Peter Green (New York: World, 1962), p. 112.

2. Foucault, "What Is Enlightenment?" trans. Catherine Porter, in Paul Rabinow, ed., *The Foucault Reader* (New York: Pantheon, 1984), p. 38.

3. Foucault, *The Order of Things*, trans. of *Les mots et les choses* by Alan Sheridan (New York: Pantheon, 1970), p. 317.

4. For a very helpful discussion of Foucault on Kantian transcendentalism, cf. Hubert Dreyfus and Paul Rabinow, *Michel Foucault: Beyond Structuralism and Hermeneutics*, 2nd ed. (Chicago: University of Chicago Press, 1983), pp. 26–43.

5. Foucault, "What Is Enlightenment?" p. 45.

6. Foucault, "Le vie: l-expérience et la science," *Revue de Metaphysique et de Morale*, 1985, p. 3. An earlier version of this essay appeared in English as the introduction to the English translation of Georges Canguilhem's *On the Normal and the Pathological* (Dordrecht: Reidel, 1978).

7. Ibid., English version, p. x.

8. Gerard Raulet, "Structuralism and Post-Structuralism: An Interview with Michel Foucault," *Telos*, 1983, p. 198.

9. Introduction to Ludwig Binswanger, *Le rêve et l'existence*, trans. Jacques Verdeaux (Paris: Desclee de Brouwer, 1954), pp. 9–128.

10. Cf. G. Bachelard, *The New Scientific Spirit*, trans. Arthur Goldhammer (Boston: Beacon, 1984), pp. 138ff. and chap. 6.

11. Cf. Bachelard, *The Philosophy of No*, trans. G. C. Waterson (New York: Orion Press, 1968), chap. 3.

12. Bachelard, *The New Scientific Spirit*, p. 31.

13. Bachelard, *Le rationalisme appliqué* (Paris: Presses Universitaires de France, 1949), chap. 7.

14. Bachelard, *Le matérialisme rationnel* (Paris: Presses Universitaires de France, 1953), p. 207.

15. Ibid., p. 219.

16. *The Philosophy of No*, p. 119.

17. Bachelard, *L'activité rationaliste de la physique contemporaine* (Paris: Presses Universitaires de France, 1951), p. 25.

18. Ibid., p. 24.

19. Canguilhem, *Études d'histoire et de philosophie des sciences* (Paris: Vrin, 1970), pp. 197–98; *Ideologie et rationalité* (Paris: Vrin, 1977), pp. 21 ff.

20. *Idéologie et rationalité*, p. 22.

21. *L'activité rationaliste de la physique contemporaine*, p. 25.

22. Ibid., p. 26.

23. *The New Scientific Spirit*, p. 45.

24. Foucault, *The Archaeology of Knowledge*, trans. Alan Sheridan (New York: Pantheon, 1972), p. 190.

25. Introduction to English translation of Canguilhem, *On the Normal and the Pathological*, p. xii.

26. *The Archaeology of Knowledge*, p. 75. Subsequent references to this book will be given in the text.

27. Cited in Dreyfus and Rabinow, p. 46, note.

28. Dreyfus and Rabinow (ibid., pp. 47–48) make the point that, unlike Searle et al., Foucault is interested in "serious speech acts" (those that have a special institutional status and autonomy) rather than "everyday speech acts." This is a helpful distinction, but I do not think it catches the central methodological difference between Foucault and speech-act theorists.

THE RAGE AGAINST REASON

Richard Bernstein

Recently, a number of philosophers including Alasdair MacIntyre, Richard Rorty, Paul Ricoeur, and Jean-François Lyotard have reminded us about the central (and problematic) role of narratives for philosophic inquiry. I say "reminded us" because narrative discourse has always been important for philosophy. Typically, every significant philosopher situates his or her own work by telling a story about what happened before he or she came along—a story that has its own heroes and villains. This is the way in which philosophers are always creating and recreating their own traditions and canons. And the stories that they tell are systematically interwoven with what they take to be their distinctive contributions. Consider Aristotle's narrative in the first book of the *Metaphysics* about the insights and blindnesses of his predecessors in grasping the multidimensional character of our scientific knowledge of causes. Or—to leap to the contemporary scene—think of the story that logical positivists have told us about the confusions and linguistic blunders of most of their predecessors—with a few bright moments of anticipation of their own radical program for reforming philosophy. Or again, there is the powerful, seductive story that Husserl

This paper, written for the conference on the Shaping of Scientific Rationality, subsequently appeared in *Philosophy and Literature*. It appears here with the agreement of the editor of that journal.

tells, where the entire history of philosophy is viewed as a teleological anticipation of the new rigorous *Wissenschaft* of transcendental phenomenology. There is a common rhetorical pattern in these narratives. They tell stories of anticipations, setbacks, and trials, but they culminate with the progressive realization of truth and reason, which is normally identified with what the philosopher/storyteller *now* sees clearly—a "truth" which his or her predecessors saw only through a glass darkly.

There is also the genre of philosophic narratives—which have become so fashionable in the nineteenth and twentieth-centuries—that dramatically reverse this pattern. They tell of relentless decline, degeneration, catastrophe, and forgetfulness. A "classic" instance of this is Nietzsche's geneological unmasking of the "history" of reason, truth, and morality, which culminates in the dominance and spread of a pernicious all-encompassing nihilism. But we also find variations of this pattern in MacIntyre's saga of the decline and degeneration of moral philosophy and moral life since the Enlightenment. And—as we shall see—this is the way in which Heidegger reads (in his "strong reading") the destiny of Western philosophy and metaphysics, which is interwoven with the history of the forgetfulness and concealment of Being.

It is not my intention to develop a typology of narrative patterns in philosophy, although I am convinced that such a typology would be extremely illuminating. Rather I want to set a context for what I will attempt to do in this paper. For I want to outline a narrative—or more accurately and modestly—a narrative sketch. Even though I will be schematic, my tale is a complex one for several reasons. First, because it is a narrative about narratives, specifically narratives which themselves relate stories about the development of reason, or what thinkers such as Weber and Habermas call "rationalization" processes.[1] Secondly, because it is a narrative that isolates different story lines, a plot and a counterplot that stand in an uneasy and unresolved tension with each other. Thirdly, because it is not one of those narratives where all the loose threads are neatly tied together at the end—or to switch metaphors, there is no grand *Aufhebung* because it is essentially an unfinished story.

In the spirit of this prologue, let me introduce the four main characters of my *first* story line and tell you what I hope to achieve. The names of the main characters are Condorcet, Weber, Adorno, and Heidegger. My aim is to confront some deeply troubling contemporary questions. For I want to understand why today there are so many "voices" screeching about Reason. Why is there a rage against Reason? What precisely is being attacked, criticized, and damned? Why is it that when "Reason" and "Rationality" are mentioned, they evoke images of domination, oppression, repression, patriarchy, sterility, violence, totality, totalitarianism, and even terror? These questions are especially poignant and perplexing when we realize that not so long ago the call to "Reason" elicited associations with autonomy, freedom, justice, equality, happiness, and peace. I not only want to understand what is happening but—even more important—what ought to be our response to the disturbing and confusing situation. Without further introduction, let me begin my tale of the battle of lightness and darkness.

1. The Dialectic of Enlightenment

In July 1793, Marie-Jean-Antoine-Nicolas Caritat, Marquis de Condorcet, under the threat of death by the Jacobins who had condemned him and declared him to be *hors la loi,* went into hiding in the house of Madame Vernet, 21 rue Servandoni, where he wrote his now famous *Esquisse d'un tableau historique des progrès de l'esprit humain* (Sketch for a historical picture of the progress of the human mind). Published posthumously in 1795 (Condorcet died the day after he was apprehended and dragged to prison in April 1794), the *Esquisse* was immediately hailed as a testament of the French Enlightenment. It was officially adopted as the philosophical manifesto of the post-Thermidorian reconstruction when the Convention voted funds to distribute copies throughout France.

It is a remarkable document. In less than two hundred pages, Condorcet sweeps through the nine stages or epochs of the history of mankind, culminating with a euphoric description of the tenth epoch, "The Future Progress of the Human Mind." "Progress" for Condorcet never simply means growth, development, and differentiation. It has a teleological normative aura—progress *toward* the indefinite perfectibility of the human species. The hero of his narrative is Reason—first manifested in philosophy, then in the natural sciences, and finally in the "moral and political sciences." Furthermore, in the course of human history, Reason gains in strength and power. With the discovery of printing, the good works of publicists and especially through public education, the full illumination of Reason spreads to all of humankind. Reason passes through difficult trials. It must triumph over the devious tactics of priests, tyrants, despots, and cunning hypocrites. But in the course of its journey through history, it gains an overwhelming momentum.

The moral of Condorcet's narrative is announced in the very beginning of the *Esquisse*. He will show us

> in the modifications that the human species has undergone, ceaselessly renewing itself through the immensity of the centuries, the path that it has made towards truth and happiness.
>
> Such observations upon what man has been and what he is today, will instruct us about the means we should employ to make certain and rapid further progress that his nature allows him still to hope for.
>
> Such is the aim of the work I have undertaken, and its result will be to show by appeal to reason and fact that nature has set no term to the perfection of human faculties; that the perfectibility of man is truly indefinite, and that the progress of this perfectibility, from now onwards independent of any power that might wish to halt it, has no other

limit than the duration of the globe upon which nature has cast us.[2]

This progress will never again be reversed as the linkage of reason, justice, virtue, equality, freedom, and happiness becomes stronger and stronger. Condorcet's history of mankind is itself teleologically oriented toward the tenth epoch—the future. He begins his "description" of "the future progress of the human *esprit*" by echoing that eighteenth-century rhetoric which was so confident that the future could be predicted on the basis of "the general laws directing the phenomena of the universe."[3] But Condorcet's "predictions" read more like utopian dreams and "hopes for the future condition of the human race."[4] There will be the eventual abolition of all forms of pernicious inequality. There will be cultural, political, and economic equality among nations and within each nation. There will be the indefinite perfection of the human faculties. Private and public happiness will prevail. Condorcet even explicitly speaks of the elimination of sexual inequality and hints (in an ambiguous manner) about stamping out racism. War will be no more, peace will eternally reign. There will even be a transformation of our biological nature. For the duration of human life will be indefinitely extended and our faculties will be strengthened.

> The time will therefore come when the sun will shine only on free men who know no other master but their reason; when tyrants and slaves, priests and their stupid or hypocritical instruments will exist only in the works of history and on the stage; and when we shall think of them only to pity their victims and their dupes; to maintain ourselves in a state of vigilance by thinking on their excesses; and to learn how to recognize and so to destroy, by force of reason, the first seeds of tyranny and superstition should they ever dare to reappear amongst us.[5]

From the perspective of the final decades of the twentieth-century with living memories of barbaric totali-

tarianism, death camps, and the ever-present danger of
nuclear cataclysm, it is difficult to resist the temptation to
read the *Esquisse,* Condorcet's testament to the future—
our present—with sardonic irony. Even such a sympa-
thetic interpreter of the Enlightenment as Peter Gay says
"the *Esquisse,* we must conclude, is as much a caricature of
the Enlightenment as its testament; it is rationalism run
riot, dominated by a simple-minded faith in science that
confuses, over and over again, the improvement of techniques with advances in virtue and happiness."[6] Condorcet's "predictions" and "hopes" have relentlessly turned
into surrealistic nightmares. Many would concur with the
judgment of Horkheimer and Adorno when they wrote:
"the Enlightenment has always aimed at liberating men
from fear and establishing their sovereignty. Yet the fully
enlightened earth radiates disaster triumphant."[7]

Yet it behooves us to remember how many of Condorcet's hopes—frequently in more modulated tones—have
animated and still animate those who come after him. For
we still hope and dream of the end of oppressive inequality, the institutionalization of freedom, and a reign of
peace. Many of us still share his faith in the potential
power of public discussion and education. Let us not forget that throughout the nineteenth and twentieth centuries, when the "social sciences" (a term already used by
Condorcet) were being developed, many of their practitioners believed—and still believe—that they provide "the
means we should employ to make certain and rapid further progress" toward human improvement and the allevi-
ation of human suffering.

A storyteller always has the license to skip historical
time, so let me abruptly jump a century and continue my
narrative with Max Weber. I want to juxtapose Weber's
chilling prognosis of our future "progress" with Condorcet's apocalyptic vision.[8] Weber is at once an heir to the
Enlightenment in his passionate commitment to reason
and the "calling" (*Beruf*) of science, and at the same time

one of its harshest and most devastating critics. Weber begins to expose what Horkheimer and Adorno call the "dialectic of Enlightenment"—the dark side of the Enlightenment, which fosters its own self-destruction. One reason why Weber is so important for my narrative is because I basically agree with Alasdair MacIntyre when he writes: "the present age in its presentation of itself is dominantly Weberian."[9]

For all of Weber's insistence on tenaciously adhering to the postulate of freedom from value judgments when engaging in empirical sociological research, his own writings (as he well knew) are filled with striking and strong judgments. Perhaps the most famous and drastic judgment—which reads like an epigraph for the twentieth century—is his "conclusion" to *The Protestant Ethic:*

> No one knows who will live in this cage in the future, or whether at the end of this tremendous development entirely new prophets will arise, or there will be a great rebirth of old ideas and ideals, or, if neither, mechanized petrifaction embellished with a sort of convulsive self-importance. For of the last stage of this cultural development, it might well be said: "Specialists without spirit, sensualists without heart; this nullity imagines that it has attained a level of civilization never before achieved."[10]

Weber was a relentless critic of the type of philosophy of history, social evolutionism, and even a "state model" theory of human development which are presupposed by Condorcet and his successors (even though recent commentators have argued there are vestiges of these in Weber's own narrative of the emergence, development, and fate of "Occidental Rationality").[11] If the Enlightenment was committed to destroying myths, superstitions, illusions, and prejudices, then Weber—in this tradition—seeks to expose and smash the mythic thought patterns of the Enlightenment itself. He scorns the very idea of teleological progress—except if we think of "progress" with

bitter irony. Freedom and republican democracy are not
the "natural" telos of human history—as Condorcet believed. On the contrary, the primary trends that characterize modernity, especially those exhibited in the
development of capitalism (and socialism) pose the greatest threat to freedom and democracy. In 1906, he wrote:

> It is utterly ridiculous to see any connection between the
> high capitalism of today—as it is now being imported into
> Russia and as it exists in America—with democracy or with
> freedom in any sense of these words. Yet this capitalism is an
> unavoidable result of our economic development. The ques-
> tion is: how are freedom and democracy in the long run at
> all possible under the domination of highly developed capi-
> talism. Freedom and democracy are only possible where the
> resolute will of a nation not to allow itself to be ruled like
> sheep is permanently alive. We are individualists and parti-
> sans of "democratic" institutions "against the stream" of
> material constellations. He who wishes to be the weathercock
> of an evolutionary trend should give up those old-fashioned
> ideals as soon as possible. The historical origin of modern
> freedom has had certain unique preconditions which will
> never repeat themselves.[12]

Not only is "mechanized petrification" an imminent
historical possibility, but freedom and democracy are en-
dangered. Weber is just as relentless in de-constructing
other key elements of Condorcet's prediction/hopes. In
one of his more nationalist pronouncements, he warns us:
"It is not peace and happiness that we shall have to hand
to our descendants, but rather the principle of eternal
struggle for the survival and higher breeding [Emporzüch-
tung] of our national species."[13] Modernity is not characterized by a universal assent to, and institutionalization
of, natural rights, but by a new polytheism of warring,
incommensurable value-commitments, by a new and violent struggle of gods and demons.

But perhaps the most severe threat to Condorcet's
hopes is Weber's challenge to the very idea that science—

as it has "progressively" developed—can tell us how we should live our lives. Condorcet never seriously questions that the sciences provide not only the *means* for human perfectibility but also reveal the *ends* to be achieved. But this myth is precisely what Weber seeks to explode. Here, too, we can detect another ironic reversal. For in Condorcet's version of Enlightenment aspirations, there is a fusion and confusion of the "is" and the "ought"—of instrumental means and normative ends. But Weber pushes to the extreme that other version of Enlightenment thinking that highlights the logical gap between the "is" and the "ought." For Kant, recognizing and insisting upon the categorical distinction of the "is" and the "ought" is the way, indeed the only way, *rationally* to ground the universal moral imperative. But for Weber, opening this abyss has the consequence of showing that there cannot be any scientific—or, more generally, *rational*—foundation for our ultimate norms.

When Weber poses the question, "What is the meaning of science?" his answer is unequivocal. He tells us:

> Tolstoy has given the simplest answer with the words: "Science is meaningless because it gives no answer to our question, the only question important for us: "What shall we do and how shall we live?" That science does not give an answer to this is indisputable. The only question that remains is the sense in which science gives "no answer," and whether or not science might yet be of some use to the one who puts the question correctly.[14]

In the background of these passionate pronouncements is a figure who, like an ominous specter, hovers not only over Weber's thought but over my own narrative: the specter of Nietzsche. There is scarcely a criticism advanced by Weber or any other critic of the Enlightenment that was not anticipated by Nietzsche—frequently in a much more succinct, sharper, and aphoristic form.

If we simply contrast the visionary optimism of Condorcet with the tragic cultural and sociological resignation
of Weber, we would leave untouched the primary question
that needs to be confronted. Why? How are we to account
for this striking and consequential difference?

An adequate answer would itself require a detailed narrative of the economic, political, and cultural developments of the century that separates them.[15] But let me
suggest that an essential clue is to be found in what stands
as *the* cherished concept of the Enlightenment—Reason.
Compared with Weber, Condorcet's understanding of reason and how it operates as a force in history is naive and
simplistic. However, Weber's frequent use of *Rationalism,
Rationalität,* and *Rationalisierung* is so complex, multidimensional and polysemous that one can well understand
why someone like Steven Lukes claims that Weber's use of
'rational' and its cognates is "irredeemably opaque and
shifting."[16] Nevertheless one of the most fruitful and
promising developments in recent Weber scholarship has
been the attempt to sort out the different meanings of
'rationality' and its cognates, the plethora of shifting dis-
tinctions that Weber introduces, and to reconstruct the
outlines of his comprehensive theory of rationalization.
Habermas is right when he suggests that using Weber's
theory of rationalization as a guideline, it is possible to
reconstruct his project as a whole.[17] Indeed, it is possible
to reveal the deep tensions and paradoxes that lie at the
very heart of his thought.

I cannot enter into the details of contemporary controversies and reinterpretations of Weber, but I do want to
highlight a few commonly accepted key points that are
important for my narrative. If we take account of Weber's
entire project for a sociology of religion and do not exclusively focus on *The Protestant Ethic,* then it becomes clear
that Weber's understanding of modern rationalization
processes is embedded within a much more comprehensive framework. He seeks to understand the specific and

peculiar rationalism of the Occident, its manifestation in the domains of culture, society, and personality, as well as the different types and developmental rhythms of rationalization.[18] Even posing the problem in this way, i.e., asking what is *distinctive* about Western rationality, indicates that Weber thinks there are forms of rationality and rationalization which are characteristic of non-Occidental cultures and societies. Furthermore, the question of Occidental rationalism needs to be further differentiated in order to comprehend the developmental patterns of Western rationality, and the distinctive forms of *modern* rationalization processes. Thus, for example, Weber's thesis about the disenchantment *(Entzauberung)* of the world (literally "de-magification") is not unique to modernity, but is a developmental process involved in the history of world religions and worldviews. This de-magification, which is itself the result of complex rationalization processes, is a necessary precondition for the appearance of modern forms of Western rationality.

Later I will return to the significance of Weber's comprehensive and internally complex scheme, but virtually all commentators on Weber agree that the type of rationalization processes that loomed so large for him in understanding the modern age are those tied to one of the four types of social action that he discriminated: purposive-rational *(zweckrational)* action.[19] "Action is purposive-rational when it is oriented to ends, means, and secondary results. This involves rationally weighing the relations of means to ends, the relations of ends to secondary consequences, and finally the relative importance of different ends."[20] This concept of purposive-rational action is the key to Weber's more complex concept of "practical rationality"—which itself is a combination of "purposive-rational" and "value-rational" action.[21] What Weber sought to show in virtually every domain of modern culture and society—including science, morality, law, politics, economics, administration, bureaucracy, even the arts—is the relentless pressure and spread of *Zweckrationalität,* which shapes every aspect of our everyday lives. It is this complex developmental process—from which there is no "turning back"—that reinforces the "iron cage" and leads to "mechanical petrification," which threatens freedom and democracy and even has the potential for undermining the very existence of the autonomous individual. This has been called "the paradox of rationalization." Albrecht Wellmer characterizes this paradox succinctly when he writes:

> through his analysis of the institutional correlates of progressive rationalization—capitalist economy, bureaucracy, and professionalized empirical science—[Weber] shows at the same time that the "rationalization" of society does not carry any utopian perspective, but is rather likely to lead to an increasing imprisonment of modern man in dehumanized systems of a new kind of increasing "reification," as Weber's disciple Lukács later would call it. The paradox, that "rationalization" connotes both emancipation and reification at the same time, remains unresolved in Weber's theory.[22]

It is this unresolved paradox that prompted Herbert Marcuse, in his own stinging critique of Weber, to remark: "It is difficult to see reason at all in the ever more solid 'shell of bondage' which is being constructed. Or is there perhaps already in Max Weber's concept of reason the irony that understands but disavows? Does he by any chance mean to say: And this you call 'reason'?"[23]

We can now understand that *if* Weber is right, *if* this is what rationality and rationalization have become in the modern world, *if* this is the "inevitable" consequence of the very type of emancipation which the Enlightenment fostered and legitimized, then we can well understand why there is a rage against reason—or, more precisely, a revulsion against what rationality has become in the contemporary world. We can see how subsequent twentieth-century critiques of the Enlightenment and its privileged forms of rationality can be understood as variations of Weberian themes. There is not only thematic continuity with Adorno's damning portrait of the "administered world," but also with Heidegger's ontologizing of Weber's paradox in his questioning of the triumph of "the will to will," the self-destructiveness of "metaphysical humanism," and the supreme danger of *Gestell* (enframing). There are also strong affinities with Foucault's microanalyses of the discursive practices of the "disciplinary society," "the carceral archipelago."

Since I still have a long way to go in my narrative, let me be briefer in introducing the two other characters of my first story line: Adorno and Heidegger. I can introduce them by highlighting a motif that runs through virtually all the types of rationality that Weber delineates—the motif of *mastery* and *control*.

Stephen Kalberg brings out this motif when he writes:

> However much they vary in content, mental processes that consciously strive to *master reality* are common to all the types of rationality. Regardless of whether they are characterized by sheer means-end calculation, the subordination of diffuse realities to values, or abstract thought processes; regardless also of whether they take place in reference to interests, formal rules and laws, values, or purely theoretical problems—all of these processes systematically confront, for Weber, social reality's endless stream of concrete occurrences, unconnected events, and punctuated happenings. In *mastering reality,* their common aim is to banish particularized perceptions by ordering them into comprehensible and "meaningful" regularities.[24] [italics added]

This is the motif that Adorno seizes upon, deepens, and seeks to explode in his negative dialectics. To capture the flavor of Adorno's paratactic and atonal style of thinking, one has to do justice to the fragile dynamic antithetical

tensions which he so tightly holds together and which are
manifested by his selective critical appropriation of
strands from Hegel, Marx, Schopenhauer, Nietzsche, Lu-
kaćs, and Benjamin. If one searches for some overall
coherent organic perspective to integrate Adorno's conflicting and even contradictory claims, one will easily be
defeated. For Adorno, fantasies of "organic wholeness"
are always regressive. The style and content of his forays
(he rejects any distinction between style and content) always aim at undermining and defeating any final
synthesis—any positive *Aufhebung*. This is one of the many
reasons why we can find in Adorno anticipations of the
deconstructive critiques of logocentrism.[25] Martin Jay suggests that we approach Adorno by applying two of his
favorite metaphors to his own style of thinking.

> The first of these is the force-field *(Kraftfeld)*, by which
> Adorno meant a relational interplay of attractions and aver-
> sions that constituted the dynamic transmutational structure
> of a complex phenomenon. The second is the constellation,
> an astronomical term Adorno borrowed from Benjamin to
> signify a juxtaposed rather than integrated cluster of chang-
> ing elements that resist reduction to a common denominator,
> essential core, or generative first principle.[26]

I want to relate these metaphors to two deep antithetical
traces which need to be juxtaposed in Adorno's own constellation.

On the one hand, the cultural and sociological pessi-
mism that flows from Weber's formulation of the paradox
of rationalization begins to look like innocent child's play
compared with Adorno's unrelentingly bleak portrait of
the "administered world" and the contemporary "culture
industry." While Weber still differentiated conflicting elements in modern forms of rationalization processes, these
are fused together in Adorno's nightmarish characterization of "instrumental rationality" gone mad. In Adorno's

allegorical representation of the Enlightenment's conception of Reason, which he labels "identity logic" or "the
philosophy of identity," repressive Reason did not arise in
the eighteenth century, but has its origins in the very beginnings of Western culture. The cunning Odysseus is the
first "bourgeois" Enlightenment figure.[27] The hidden
structure of the "identity logic," which is expressed in
conceptual form in the beginnings of Western philosophy
and reaches its culmination in Hegel, who is tempted by
the lure of the System, turns out to be the will to mastery
and control. This is just the motif that runs through Weber's characterizations of the types of rationality. But for
Adorno this identity rationality always seeks to deny, re-
press, and violate otherness, difference, and particularity.
This form of reason—when unmasked—is intrinsically
domination *(Herrschaft)*; the domination and control over
nature inexorably turns into the domination of men over
men (and indeed men over women) and culminates in
sadistic-masochistic self-repression and self-mutilation.
The hidden "logic" of Enlightenment reason is violently
repressive; it is totalitarian. Lyotard echoes Adorno (and
reveals how Adorno anticipated recent postmodern controversies) when he concludes his essay, "What is
postmodernism?" by declaring:

> The nineteenth and twentieth centuries have given us as
> much terror as we can take. We have paid a high enough
> price for the nostalgia of the whole and the one, for the
> reconciliation of the concept and the sensible, of the trans-
> parent and the communicable experience. Under the general
> demand for slackening and for appeasement we can hear the
> mutterings of the desire for a return of terror, for the realiza-
> tion of the fantasy to seize reality. The answer is: Let us wage
> war on totality; let us be witnesses of the unpresentable; let
> us activate the differences and save the honor of the name.[28]

On the other hand, however, despite Adorno's scathing
exposure of the dark sado-masochistic side of the Enlight-

enment, whose legacy is epitomized in that single horrible name, *Auschwitz,* he is still an heir—albeit in a different way than Weber—to Enlightenment aspirations. Even in the introduction to the coauthored *Dialectic of Enlightenment,* which many have read as Horkheimer's and Adorno's most pessimistic work, the authors declare:

> We are wholly convinced—and therein lies our *petitio principii*—that social freedom is inseparable from enlightened thought. Nevertheless, we believe that we have just as clearly recognized that the notion of this very way of thinking, no less than the actual historic forms—the social institutions—with which it is interwoven, already contains the seed of the reversal universally apparent today.[29]

Initially, what seems so paradoxical—indeed aporetic—is the way in which Adorno self-consciously affirms the wildest utopian dreams of the Enlightenment—its *promesse de bonheur* and the end of all human suffering. The utopian hopes of the Enlightenment which Weber scorned as "unrealistic" and "irrational" are hyperbolically affirmed by Adorno. And happiness for Adorno is not a pale public *eudaimonia* or private well-being, but an aestheticized, unrepressed sensuous gratification and ease. Adorno—in his own mimesis of redemption—holds up before us the vision of a nonantagonistic, nonhierarchical, nonviolent, and nonrepressive society.

Adorno presses to radical extremes these two elements into a new constellation—the unlimited *dynamis* of the *promesse de bonheur* and the devastating distorting power of "identity logic"—not to foster the illusion of a grand *Aufhebung* but rather to crack open encrusted repressive social reality.

> The only philosophy which can be responsibly practised in the face of despair is the attempt to contemplate all things as they would present themselves from the standpoint of redemption. Knowledge has no light but that shed on the world by redemption: all else is reconstruction, mere tech-

nique. Perspectives must be fashioned that displace and es-
trange the world, reveal it to be, with its rifts and crevices, as
indigent and distorted as it will appear one day in the messi-
anic light.[30]

Adorno bequeaths us a whole cluster of aporias. But
the fragile and taut thread that still connects Adorno with
the deepest strata of the Enlightenment's *promesse de bonheur* is broken by Heidegger. For all of Adorno's explicit
scorn for Heidegger and despite his barbed condemnation of the "jargon of authenticity," Adorno provides the
bridge—or to use Nietzsche's phrase, "the tightrope"—to
Heidegger's ontological rendering of the history and destiny of logos and reason which culminates in metaphysical
humanity's blindness and forgetfulness of the silent call of
Being. In Heidegger's fateful strong reading of the "history of Being," which is already foreshadowed in *Being and
Time* but becomes more and more pronounced in his
"middle" and "late" writings, we find a thematic affinity
with Adorno's own claim that the seeds of "identity logic"
with its hidden will to mastery are to be found in the very
origins of Western rationality. For Heidegger, there is a
direct continuity—a single story to be told of the playing
out of "prefigured possibilities" from Plato to Nietzsche,
the last metaphysical thinker. "With Nietzsche's metaphysics, philosophy is completed."[31] The very idea of *animal
rationale*—metaphysical man—is emblematic of the obliv-
ion of Being. The mood *(Stimmung)* of Heidegger's "history of reason" and more encompassing "history of
being" is manifest when he writes:

> The decline of the truth of beings occurs necessarily, and
> indeed as the completion of metaphysics. The decline occurs
> through the collapse of the world characterized by metaphy-
> sics, and at the same time through the desolation of the earth
> stemming from metaphysics. Collapse and desolation find
> their adequate occurrence in the fact that metaphysical man,
> the *animal rationale*, gets fixed as the laboring animal. This

rigidification confirms the most extreme blindness to the ob-
livion of *Being*. But man wills *himself* as the volunteer of the
will to will, for which all truth becomes error which it needs
in order to be able to guarantee for itself the illusion that the
will to will can will nothing other than empty nothingness, in
the face of which it asserts itself without being able to know
its own completed nullity.[32]

The specter of Nietzsche which haunts Weber and
Adorno now takes on an ominous hyper-reality. It is difficult to resist the conclusion that we are—as Gadamer says
of Heidegger—living in "the 'cosmic night' of the 'forgetfulness of being,' the nihilism that Nietzsche prophe-
sied."[33] Heidegger does tantalize us by suggesting that we
may yet overcome *(überwindung)* metaphysics, we may still
be "saved" from the supreme danger, the essence of technology, *Gestell,* by a new/old event/appropriation of thinking and poetic building.[34] But we can bring to a closure
this story line by recalling Condorcet's posthumously published testament to the tenth epoch of mankind and juxta-
posing it with Heidegger's own posthumously published
testament:

> Philosophy will not be able to effect any direct transforma-
> tion on the present state of the world. This is true not only of
> philosophy but of any simply human contemplation and
> striving. Only a god can save us now. We can only through
> thinking and writing prepare to be prepared for the manifes-
> tation of god, or the absence of god as things go downhill all
> of the way.[35]

After Heidegger, it would seem that all talk of
humanism—*human* freedom, happiness, and emancipation—has become a mockery. If this were the history
and fate of Western rationality, then the rage against
reason would be eminently "reasonable." For it would
seem to be impossible to resist the "conclusion" that
the working through of the "prefigured possibilities" of

logos ineluctably results not in illumination and Enlightenment, but the cosmic black night of nihilism.

2. Dialectical Rationality

Thus far I have followed my first story line without
hinting at my other story line or counterplot. I want to
introduce it *in medias res* by returning to Weber and specifically to Habermas's recent critical response to Weber.
The "Habermas" who is a character in this narrative is
not the Habermas who has recently given birth to what
may strike many as a new form of academic scholasticism
or a new culture industry. I share with William James the
conviction that "under all the technical verbiage in which
the ingenious intellect of man envelops them, are just so
many visions, modes of feeling the whole push, and seeing the whole drift of life, forced on one by one's total
character and experience."[36] Without underestimating the
importance of "technical verbiage," I believe that Habermas himself shares this conviction. Another way of putting this is to say that one mark of an original thinker is to
gravitate toward a single thought or intuition which is explored, amplified, and probed over and over again. Indeed, if asked, as Rabbi Hillel once was in another
situation, to sum up this single dominant thought in a
phrase, I can think of no better phrase than one recently
used by Habermas in an interview. He speaks of "the
conviction that a humane collective life depends on the
vulnerable forms of innovation-bearing, reciprocal and
unforcedly egalitarian everyday communication." (Inci-
dently, and the significance of this will soon become evident, Habermas makes this remark in speaking about the
"intuition" that he sees as linking him with Richard
Rorty—an "intuition" which "Rorty retains from the
pragmatist inheritance.")[37]

But let me turn to Habermas's encounter with Weber to show how central this intuition is for him and for my own narrative. In the second chapter of *The Theory of Communicative Action,* Habermas—in what is virtually an independent monograph on Max Weber—seeks to reconstruct Weber's theory of rationalization. The details of this reconstruction are complex, challenging, and controversial—worthy of careful systematic analysis. But, informally, I think it is fair and accurate to say that what motivates Habermas throughout his critical reconstruction is almost a "gut feeling" that something has gone desperately wrong in Weber's analysis of modern rationalization processes, that something is askew and distorted. In part, Habermas seeks to show deep tensions within Weber's own project and that the "conclusions" Weber draws about the "iron cage" of modern forms of rationalization processes do not follow from, and indeed are even inconsistent with, his more comprehensive understanding of different types of rationalization processes. But perhaps even more important, Habermas wants to understand what led Weber astray—what led him to the seductive story of the ineluctable triumph of *Zweckrationalität,* which has so pervasively influenced subsequent critiques of the Enlightenment legacy. To put a very complex story into a nutshell, Habermas argues that despite all of Weber's enduring insights, the primary "conceptual bottleneck"[38] that prejudices Weber's understanding of societal rationalization can be traced back to his limited understanding of social action. By giving such prominence to purposive-rational *(zweckrational)* action, Weber slights the distinctiveness, centrality, and indeed (for Habermas) the primacy of communicative action—the type of action manifested most clearly in speech that is oriented not to success, but to mutual understanding. Whereas the rationalization of purposive-rational action, as Weber well understood, involves "the empirical efficiency of technical means and the consistency of choice

between suitable means," the rationalization of communicative action is radically different:[39]

> Rationalization here means extirpating those relations of
> force that are inconspicuously set in the very structures of
> communication and that prevent conscious settlement of
> conflicts, and consensual regulation of conflicts, by means of
> intrapsychic as well as interpersonal communicative barri-
> ers. Rationalization means overcoming such systematically
> distorted communication in which the action-supporting
> consensus concerning the reciprocally raised validity
> claims—especially consensus concerning the truthfulness of
> intentional expressions and the rightness of underlying
> norms—can be sustained in appearance only, that is, coun-
> terfactually.[40]

This reference to "counterfactuality" is extremely important for several reasons. First, because Habermas argues that there is no historical necessity involved in
specific forms and domination of *Zweckrationalität* in the
modern world. (Weber, who himself eschewed all appeals
to "historical necessity," might even agree.) In this respect
Habermas is also strongly opposing those tendencies in
Adorno and Heidegger which suggest the ineluctable destiny of the working out of "identity logic" or the pernicious domination of *Gestell.*

Secondly, the conceptual shift that Habermas introduces, dissolves the so-called paradox of rationalization.
Wellmer points this out clearly when he writes:

> Habermas objects that this paradox of rationalization does
> *not* express an internal *logic* (or dialectic) of modern rational-
> ization processes; it is strictly speaking not a paradox *of ra-
> tionalization. . . .* rather it would be more adequate to speak
> of a "selective" process of rationalization, where the selective
> character of this process may be explained by the peculiar
> restrictions put upon communicative rationalization by the
> boundary conditions and the dynamics of a capitalist system
> of production.[41]

But what is most important about Habermas's conceptual reorientation, his paradigm shift, is that it enables us
to grasp more perspicuously what Weber described without accepting his aporetic prophesies about the "iron
cage" which imprisons us. The "paradox of rationalization" is now reinterpreted as the powerful, and indeed the
dominant, tendency in the modern world toward the de-
formation of the life-world (with its distinctive communicative rationality) by the distorting pressures of systems of
purposive rationality. This conceptual shift does not mean
that Habermas is any more optimistic than Weber was
about our future prospects. Rather this conceptual shift
alters our *theoretical* understanding of the dynamic and
conflicting rationalization processes of modernity and also
our *practical* evaluation of new social movements. For we
can evaluate them as defensive strategies for protecting
and furthering the integrity of an undistorted life-world.
This is the perspective from which Habermas himself
evaluates the women's movement—a movement which he
suggests may potentially be the most radical contemporary
social movement.

I cannot go into the exposition and critique of the complex way in which Habermas seeks to explain and defend
these striking claims.[42] My primary intention here is to
make use of Habermas's insights to expand my narrative
by pursuing two lines of affinity. The first goes back to
that "intuition" which Habermas says links him with the
pragmatic tradition. For Habermas is profoundly right in
recognizing that the basic intuition or judgment that
stands at the center of his own vision is also central to the
pragmatic tradition. Both share an understanding of rationality as intrinsically dialogical and communicative.
And both pursue the ethical and political consequences of
this form of rationality and rationalization. It was Peirce
who first developed the logical backbone of this thought
in his idea of the fundamental character of a self-
corrective critical community of inquirers without any ab-

solute beginning points or finalities. It was Dewey who
argued that the very idea of such a community, when
pressed to its logical extreme, entails the moral ideal of a
democratic community where "the task of democracy is
forever that of creation of a freer and more human experience in which all share and to which all contribute."[43]
Dewey, no less than Habermas, knew how vulnerable such
"innovation-bearing, reciprocal and unforcedly egalitarian" communities can be in the face of all those tendencies in the contemporary world which seek to undermine,
crush, and deform communicative rationality. And it was
Mead who saw that the linkage of dialogic communicative
rationality and the institutionalization of democratic
forms of life requires a new understanding of the genesis
and development of practical sociality—what one recent
perceptive commentator aptly calls Mead's theory of practical intersubjectivity.[44] I vividly recall my own shock of
recognition when I first started reading Habermas in the
1960s. For I realized that he, who was primarily intellectually shaped by the German tradition from Kant through
Hegel to Marx and by his creative appropriation from the
Frankfurt School, was moving closer and closer to the
central themes of the American pragmatic tradition.[45]

To amplify my counterplot, it is helpful to pursue another line of affinity which takes us in a very different
direction—Habermas's affinity with Gadamer and with
the tradition of hermeneutics and practical philosophy
that Gadamer has helped to revivify. I am certainly aware
of the consequential differences between Habermas and
Gadamer in the understanding of the character and pre-
conditions required for dialogic rationality.[46] Habermas's
broad sociological concerns are completely alien to Gadamer's ontological hermeneutics and to Gadamer's own
self-identification with the tradition of German Romanticism.[47] But in this context, I want to focus on the common
ground—what Gadamer would call *die Sache Selbst*— that
binds them together in the twists and turns of their ongo-

ing debate, which has lasted now for more than twenty
years.

By an independent pathway—which owes a great deal
to his critical appropriation of themes in Plato, Aristotle,
Hegel, and Heidegger—Gadamer in his ontological version of hermeneutics has been arguing that our ontological condition, our very being-in-the-world, is to *be*
dialogical *beings*. We find in Gadamer one of the subtlest
and most sensitive phenomenological analyses of the internal play and essential openness of dialogue.[48] His very
practice of philosophy is constantly showing us what authentic dialogue involves. The main reason, however, why
Gadamer is so important for my second story line is that
he shows us a way of recognizing that *die Sache Selbst* of
Habermas's understanding of communicative rationality
is *not* a discovery of the twentieth century. Rather it is one
of the oldest and most persistent—albeit subterranean—
themes in Western philosophy. Although Gadamer's great
hero is the Plato of the Socratic/Platonic Dialogues, he
finds traces of this theme already in Heraclitus. Even
when he integrates Aristotle into his narrative of the history of Western rationality, he highlights the communicative aspects of *phronesis* by showing us how it presupposes
and fosters the civic virtue of friendship. Gadamer is not
motivated by a nostalgia for a "golden" past. He wants to
redeem those "truths" in the tradition of philosophy that
can enable us to understand and critically evaluate the
deformations of modern scientific technological culture
with its deification of technical expertise. Just as I rashly
attempted to sum up Habermas in a sentence, let me try
to do the same for Gadamer. For I believe that Gadamer's
entire corpus can be read as an invitation to join him in
the rediscovery and redemption of the richness and concreteness of our dialogical being-in-the-world.

By now it should be clear that I could have told my
narrative as a series of footnotes to Plato. The first plot
can be traced back to the "construction" of the metaphys-

ical Plato, with his two-world theory, his denigration of corporeality, and his celebration of eternal immutable forms which are the erotic telos of *dianoia* and *noesis*. This Plato (sometimes called "Platonism") is the villain to whom we can trace back everything that has subsequently gone wrong with Western rationality. This is the Plato who is attacked and "de-constructed" by Nietzsche, Heidegger, Derrida, and Rorty.

But there is the "other" Plato—the progenitor of my second story line—who is the great defender of the *spoken* and *written* dialogue—which is always open to novel turns and which knows no finality.[49] Much of what Habermas, the pragmatists, and even Gadamer have explored may well be understood as a commentary on the claim that it is a fiction to think "Socratic dialogue is possible everywhere and at any time."[50] All are concerned, although in very different ways, with probing the conditions required to foster the concrete embodiment of this "ideal fiction." John Dewey, who some might think an unlikely advocate of a back-to-Plato movement, put the point beautifully when he wrote:

> Nothing could be more helpful to present philosophizing than a "Back to Plato" movement; but it would have to be a back to the dramatic, restless, cooperatively inquiring Plato of the Dialogues, trying one mode of attack after another to see what it might yield; back to the Plato whose highest flight of metaphysics always terminated with a social and practical turn, and not to the artificial Plato constructed by unimaginative commentators who treat him as the original university professor.[51]

3. *The New Constellation*

My narrative is not yet finished because the story I am telling is still unfolding. At the beginning I warned that this is not one of those stories where all the loose threads are neatly tied up at the "end." Let me once again recall

Benjamin's and Adorno's metaphor of a "constellation," and suggest that today we find ourselves in a new constellation. In our "post-era"—whether we label it postmodernist, poststructuralist, postmetaphysical (or any of the dozen or so "post" appellations which are being bandied around), there are those who are telling us that the very idea of dialogue and communicative rationality belong to the dustbin of the now discredited history of Western rationality and metaphysics. These ideas and ideals are part of the now exhausted metaphysics of presence, logocentrism, phonocentrism, ethnocentrism, and phallocentrism which comprise the violent history of the West. To the extent that this has now become the fashionable, "sophisticated" skeptical claim of the movement called deconstruction, I think it is radically mistaken. Furthermore—and this may surprise you—I do not for a moment think that this is what Derrida, with whom the name "deconstruction" has become entangled, is really telling and showing us. Since I am coming to the end of my tale, let me state briefly and forcefully what I take to be—to use an old-fashioned term—the *truths* to be appropriated from so-called postmodern debates. Isaiah Berlin once commented that "the history of thought and culture is, as Hegel showed with great brilliance, a changing pattern of great liberating ideas which inevitably turn into suffocating straitjackets, and so stimulate their own destruction by new emancipating, and at the same time, enslaving conceptions."[52]

What such thinkers as Derrida—and in a very different manner, Foucault—have shown us is that such ideas as authentic dialogue, community, communication, and communicative rationality *can* potentially—and indeed *have* in the past—become "suffocating straitjackets," and "enslaving conceptions." This is already anticipated by Benjamin's and Adorno's deep suspicion of what "communication" has become in an administered world: little more than the technological exchange of information to be utilized—input and output of "data." We need only listen to the political rhetoric of the leaders of the great super-powers to hear what 'dialogue' means today—a form of skillful manipulation which seeks to obtain a greater military advantage.

But there are more subtle, unobtrusive, but even more pernicious dangers that need to be unmasked and revealed. There can be no dialogue, no communication unless beliefs, values, commitments, and even emotions and passions are shared in common. Furthermore, I agree with Gadamer and MacIntyre that dialogic communication presupposes moral virtues—a certain "good will"—at least in the willingness to really listen, to seek to understand what is genuinely other, different, alien, and the courage to risk one's more cherished prejudgments. But too frequently this commonality is not really shared, it is *violently* imposed. A false "we" is projected. As I read Derrida, few contemporary writers equal him in his sensitivity and alertness to the multifarious ways in which the "history of the West"—even in its institutionalization of communicative practices—has always tended to silence differences, to exclude outsiders and exiles, those who live on the margins. The so-called "conversation of mankind" has been just that—a conversation of *mankind*, primarily white mankind. This is one of the many *good* reasons why Derrida "speaks" to those who have felt the pain and suffering of being excluded by the prevailing hierarchies embedded in the text called "the history of the West"—whether they be women, blacks, or others bludgeoned by exclusionary tactics. Even Derrida's deconstructive inversion and reinscription of speech and writing can be read as a warning against the nostalgic belief that face-to-face spoken language is sufficient to guarantee communication. He teaches us how much can go wrong—even tragically wrong—in the folds of communication.[53]

As for Foucault—at his best—he shows us that if we take a cold, hard look at the discursive practices that un-

derlie so much of modern "humanism" and the human
sciences, we discover power/knowledge complexes that be-
lie what their ideologues profess. In novel ways Foucault
shows us the truth of Benjamin's claim, "There is no document of civilization which is not at the same time a
document of barbarism."[54] Sometimes what is required to
communicate—to establish a reciprocal "we"—is rupture
and break—a *refusal* to accept the common ground laid
down by the "other." It is extremely easy to pay lip service
to recognizing and respecting genuine plurality, difference, otherness, but nothing perhaps is more difficult
than to achieve this in practice—and such practice is *never*
completely stable or permanent. It is a self-deceptive illusion to think that the "other" can always be heard in a
friendly dialogue.

I would go further and argue that Derrida, Foucault,
Lyotard, and many others are important not only for their
"negative" cautious moral skepticism in warning us of the
hidden dangers of "false" consensus, dialogue, community, a "false" we, but that when we think through what
they are saying, when we try to make sense of their own
moral passion, we are led back to the fragile, but persistent "ideal" of dialogical communicative rationality—an
ideal which is more often betrayed than honored.

We must learn again and again to hear what Weber,
Adorno, Heidegger, and their successors are telling and
showing us. But we must resist the temptation to be seduced by "arguments" of necessity, destiny, and ineluctable decline. We must resist those essentialist stories of
the history of Western rationality that see it as *only* ending
in hidden forms of violence and despairing nihilism. For
then we would surely be enclosed in the darkness of forgetfulness and betrayal. Let us not forget that "communicative reason operates in history as an avenging force,"[55]
and that the claim to reason has a "stubbornly transcending power, because it is renewed with each act of unconstrained understanding" and with "each moment of living

together in solidarity." But never before has *this* claim to communicative reason been so threatened from so many different directions. A *practical* commitment to the avenging *energeia* of communicative reason is the basis— perhaps the only honest basis—for hope.

NOTES

1. The expression "rationalization" can be misleading because, in an Anglo-American context, it typically connotes a false, misleading, and distortive justification. We speak, for example, of a rationalization of hidden motives. But the expression, as used by Weber, Habermas, and others influenced by the German sociological tradition, does not have a pejorative connotation. It refers to a developmental process by which a type of rationality increases over time. Thus for Weber an increase in the efficiency of bureaucratic administration or the development of empirical science would both be understood as rationalization processes.

2. Antoine-Nicolas de Condorcet, *Sketch for a Historical Picture of the Progress of the Human Mind,* trans. June Barraclough (London: Weidenfeld & Nicolson, 1955), p. 4.

3. Ibid., p. 173.

4. Ibid.

5. Ibid., p. 179.

6. Peter Gay, *The Enlightenment: An Interpretation* (New York: Alfred A. Knopf, 1969), vol. 2, p. 122.

7. Max Horkheimer and Theodor W. Adorno, *Dialectic of Enlightenment* (New York: Herder & Herder, 1972), p. 3.

8. Louis-Gabriel-Ambroise Bonald hailed the *Esquisse* as the "apocalypse of the new gospel." See Keith Michael Baker, *Condorcet: From Natural Philosophy to Social Mathematics* (Chicago: University of Chicago Press, 1975), p. 393.

9. Alasdair MacIntyre, *After Virtue* (Notre Dame, Ind.: University of Notre Dame Press, 1981), p. 108.

10. Max Weber, *The Protestant Ethic and the Spirit of Capitalism,* trans. Talcott Parsons (New York: Scribner's, 1958), p. 182.

11. See Wolfgang Schluchter, *The Rise of Western Rationalism: Max Weber's Developmental History,* trans. Guenther Roth (Berkeley: University of California Press, 1981), and Jürgen Habermas, *The Theory of Communicative Action,* trans. Thomas McCarthy, 2 vols. (Boston: Beacon, 1981), vol. 1, chap. 2, "Max Weber's Theory of Rationalization."

12. *Archiv für Social wissenschaft und Sozialpolitik,* vol. 12, no. 1, pp. 347 ff., cited in H. H. Gerth and C. Wright Mills, *From Max Weber: Essays in Sociology* (New York: Oxford University Press, 1946), p. 71.

13. Cited by Wolfgang J. Mommsen, *The Age of Bureaucracy* (New York: Harper & Row, 1974), p. 30.

14. Max Weber, "Science as a Vocation," in *From Max Weber: Essays in Sociology,* p. 143.

15. See Habermas's account in his analysis of "Max Weber's Theory of Rationalization."

16. Steven Lukes, "Some Problems about Rationality," in Bryan Wilson, ed., *Rationality* (New York: Harper & Row, 1971), p. 207.

17. Jürgen Habermas, *A Theory of Communicative Action* 1:143.

18. See the Introduction to *The Protestant Ethic,* which is a translation of the *Vorbemerkung* to his studies in the sociology of religion.

19. The other three types of social action are affectual, traditional, and value-rational action. See Stephen Kalberg's discussion of the types of social action and the types of rationality in "Max Weber's Types of Rationality: Cornerstones for the Analysis of Rationalization Processes," *American Journal of Sociology* 85 (1980): 1145–79. Kalberg distinguishes four types of rationality: theoretical, practical, substantive, and formal.

20. Max Weber, *Economy and Society,* trans. Guenther Roth and Claus Wittich, 2 vols. (Berkeley: University of California Press, 1978) 1:26. The translation is slightly modified.

21. See Habermas's "reconstruction" of the concept of practical rationality in *The Theory of Communicative Action* 1:168 ff.

22. Albrecht Wellmer, "Reason, Utopia, and the *Dialectic of Enlightenment*" in *Habermas and Modernity,* ed. Richard J. Bernstein (Cambridge: M.I.T. Press, 1985), p. 41.

23. Herbert Marcuse, "Industrialization and Capitalism in the Work of Max Weber," in *Negations,* trans. Jeremy J. Shapiro (Boston: Beacon, 1968), pp. 225–26.

24. Stephen Kalberg, "Max Weber's Types of Rationality," pp. 1159–60.

25. See Martin Jay's discussion of Adorno's anticipations of deconstructionism in *Adorno* (Cambridge, Mass.: Harvard University Press, 1984), p. 21.

26. Martin Jay, *Adorno,* pp. 14–15.

27. See Max Horkheimer and Theodor W. Adorno, *Dialectic of Enlightenment.* I am focusing here on Adorno's thought, leaving aside some of the important differences between Horkheimer and Adorno. In the preface to the new English (1972) edition of *Dialectic of Enlightenment,* the authors write: "No outsider will find it easy to discern how

far we are both responsible for every sentence. We jointly dictated lengthy sections; and the vital principle of the *Dialectic* is the tension between the two intellectual temperaments conjoined in it," p. ix.

28. Jean-François Lyotard, *The Post-Modern Condition: A Report on Knowledge,* trans. G. Bennington and B. Massumi (Minneapolis: University of Minnesota Press, 1984), pp. 81–82.

29. *Dialectic of Enlightenment,* p. xiii.

30. Theodor W. Adorno, *Minima Moralia,* trans E. F. N. Jephcott (London: New Left Bookstore, 1974), p. 247.

31. Martin Heidegger, "Overcoming Metaphysics" in *The End of Philosophy,* trans. Joan Stambaugh (New York: Harper & Row, 1973), p. 95.

32. "Overcoming Metaphysics," p. 86.

33. Hans-Georg Gadamer, *Truth and Method,* trans. G. Barden and J. Cumming (New York: Seabury, 1975), p. xxv.

34. See "The Question Concerning Technology," especially Heidegger's interpretation of Hölderlin's lines "But where danger is, grows / The saving power also" in Martin Heidegger, *Basic Writings,* ed. and trans. David F. Krell (New York: Harper & Row, 1977), pp. 283–318. See my critical discussion, "Heidegger on Humanism" in *Philosophical Profiles* (Philadelphia: University of Pennsylvania Press, 1986).

35. Martin Heidegger, "Only a God Can Save Us Now," trans. D. Schendler, *Graduate Faculty Philosophy Journal* (New School for Social Research) 6 (1977), 5–27; see p. 16.

36. William James, *A Pluralistic Universe* (Cambridge: Harvard University Press, 1977), pp. 14–15.

37. Jürgen Habermas, "A Philosophico-Political Profile" in *New Left Review,* no. 151 (May/June 1985), p. 12. Habermas makes this remark in response to the question "And could you explain the discrepancy between your condemnation of poststructuralism, and your comparatively friendly reception of the work of Richard Rorty, which provides parallels to, and has in some cases been directly influenced by, poststructuralist themes?" Habermas replies: "As far as Richard Rorty is concerned, I am no less critical of his contextualist position. But at least he does not climb aboard the 'anti-humanist' bandwagon, whose trail leads back in Germany to figures as politically unambiguous as Heidegger and Gehlen. Rorty retains from the pragmatist inheritance, which in many, though not all, respects he unjustly claims for himself, an intuition which links us together—the conviction that a humane collective life depends on the vulnerable forms of innovation-bearing, reciprocal and unforcedly egalitarian everyday communication."

38. *The Theory of Communicative Action* 1:270 ff.

39. Jürgen Habermas, *Communication and the Evolution of Society* (Boston: Beacon, 1979), p. 117.

40. *Communication and the Evaluation of Society,* pp. 119–20.

41. Albrecht Wellmer, "Reason, Utopia, and the *Dialectic of Enlightenment,*" p. 56.

42. For a critical discussion of some aspects of Habermas, see my discussion of his work in *The Restructuring of Social and Political Theory* (Philadelphia: University of Pennsylvania Press, 1978), and *Beyond Objectivism and Relativism: Science, Hermeneutics and Praxis* (Philadelphia: University of Pennsylvania Press, 1983).

43. John Dewey, "Creative Democracy—The Task Before Us," reprinted in M. Fisch, ed., *Classic American Philosophers* (New York: Appleton-Century-Crofts, 1951), p. 394. See my discussion of Peirce and Dewey in *Praxis and Action* (Philadelphia: University of Pennsylvania Press, 1971).

44. Hans Joas, *G. H. Mead: A Contemporary Re-examination of His Thought* (Cambridge: M.I.T. Press, 1985).

45. See Habermas's comments about Peirce, Dewey, and Mead in "A Philosophico-Political Profile," pp. 76–77.

46. See my discussion of Gadamer and Habermas in *Beyond Objectivism and Relativism.*

47. See "A Letter by Professor Hans-Georg Gadamer," included as an Appendix in *Beyond Objectivism and Relativism.*

48. See Part Three, "From Hermeneutics to Praxis," *Beyond Objectivism and Relativism.*

49. For an illuminating interpretation of Plato's *Phaedrus* which shows how Plato defends "a philosophic art of writing," see Ronna Burger, *Plato's Phaedrus: A Defense of a Philosophic Art of Writing* (University, Ala.: University of Alabama Press, 1980). Compare this with Jacques Derrida "Plato's Pharmacy," in *Dissemination* (Chicago: University of Chicago Press, 1981).

50. Jürgen Habermas, *Knowledge and Human Interests,* trans. Jeremy J. Shapiro (Boston: Beacon, 1971), p. 314.

51. John Dewey, "From Absolutism to Experimentalism" in *John Dewey: On Experience, Nature, and Freedom,* ed. Richard J. Bernstein (New York: Library of Liberal Arts, 1960), p. 13.

52. Isaiah Berlin, "Does Political Theory Still Exist?" in *Philosophy, Politics, and Society* (2nd series), ed. Peter Laslett and W. G. Runciman (Oxford: Basil Blackwell, 1962), p. 17.

53. Henry Staten eloquently expresses this, when in summing up his judgment of the Searle-Derrida debate, he writes: "Perhaps what we have in this debate is a conflict between Anglo-American clean-mindedness or sincerity and a more archaic moral rigor that insists on

reminding us of the residue of darkness in man's intention. If there is any skepticism in Derrida, it is a moral, not an epistemological, skepticism—not a doubt about the possibility of morality but about an idealized picture of sincerity that takes insufficient account of the windings and twistings of fear and desire, weakness and lust, sadism and masochism and the will to power, in the mind of even the most sincere man." *Wittgenstein and Derrida* (Lincoln: University of Nebraska Press, 1984) pp. 126–27.

54. Walter Benjamin, "Theses on the Philosophy of History," in *Illuminations,* ed. Hannah Arendt (New York: Schocken, 1969), p. 256.

55. Jürgen Habermas, "A Reply to My Critics" in *Habermas: Critical Debates,* ed. John B. Thompson and David Held (London: Macmillan, 1982), p. 227. See my elaboration of this theme in the Introduction to *Habermas and Modernity.*

PANEL DISCUSSION: CONSTRUCTION AND CONSTRAINT

The participants in this discussion, which was chaired by Philip Quinn, were Richard Bernstein, Gary Gutting, Richard Rorty, Ernan McMullin, Carl Hempel, Mary Hesse, Thomas McCarthy, and Alasdair MacIntyre.

Gutting: I thought I might make a few opening comments about the overall sense I got from an extremely challenging and fruitful two days of discussion. One way to get such a sense was to reflect on the title of the conference itself and on a variant of it. When I first saw the title: "The shaping of scientific rationality," I thought that it contrasted with what might well have been the title of such a conference were it to have been held fifteen or twenty years ago. It might then have been called "The *nature* of scientific rationality;" a conference at that time on that topic might have asked whether or not one could talk about scientific rationality in the kinds of historical or social terms that would allow one to speak of a "shaping." But this conference began from the assumption embodied in its title that there is a historical and social character to scientific rationality, an assumption that would surely not have been accepted only a short while ago. We have now come to take seriously the historical,

---

This text is an edited and slightly rearranged version of the panel discussion, taken from tape. The transcription was made by Ernan McMullin.

and perhaps the social, character of scientific rationality.

A second thought about the title was prompted by the heading in last night's newspaper on what was otherwise an accurate account of the conference. It was one that the organizer of the conference would assuredly not have accepted: "The shaping of scientific *reality*." The misprint was in fact a creative one. It suggests a question that has been central at the meeting: can we accept the historical and social character of scientific reasoning, as we all seem to be doing, and still maintain the objectivity and the truth of science? If we admit that the rationality of science is a historical and social phenomenon in an essential way, then what happens to the vaunted claim of science that it gets at a fixed reality? The newspaper title suggests that maybe there isn't a fixed and objective reality there to begin with but that it is created. That is the problem that many of the speakers have been struggling with, and the struggle has been for what people regard as the high middle ground, enabling them to say: yes, it's historical and social but it's still true and objective. That's where the specter of relativism begins to appear, as it did especially in the last three papers [Hesse, McCarthy, MacIntyre]. My view is that although there were strong efforts at exorcism of the specter—culminating in Alasdair's dramatic "go away!"—the specter has not really been exorcised.

A further question did come up, one that could furnish the topic for another conference. Some of the things I said about Foucault suggest it; some of the things that Dick Rorty said suggest it in another way. If the question we mainly dealt with was: given the historical/social character of scientific reasoning, how can we guarantee objectivity and truth in science?, the further question raised by Foucault and Rorty is: what's so great about objectivity and truth? That question can be raised in several different ways. Rorty raises it by asking: do objectivity and truth, ideal notions as they are, make *sense* as philosophical ideals or do they represent a blind alley in the path of inquiry? Foucault raises the question in a different way by asking: are these ideals even perhaps dangerous to human freedom? If you take *that* point of view, and it certainly has come up in our discussions, you might think of a further conference that would ask about the *threat* of scientific rationality.

Bernstein: I will use my quota of time to challenge Ernan McMullin and Alasdair MacIntyre. Despite all your pretense of being moderate and being in the center, Ernan, I want to suggest that you are really, in your scheme of things, a leftist *manqué*. I find a disparity in your paper. Let me describe it in several ways and then justify my claim. Using Hegel's distinction between what is intended and what is actually said, it is perfectly clear what you intended, but if one follows your argument through, I am not clear that there's any difference that makes a difference between you and Shapere. Not that I like Shapere's view, but I want you to face up to this fact. Or to use a slightly different rhetorical device, one that Feyerabend uses in *Against Method* when he discusses Lakatos, and claims that Lakatos too is really an "epistemological anarchist"—that the only difference between them is a matter of verbal ornament. That is what I want to accuse you too of in your apparent disagreement with Shapere.

To justify this, let me take up an example that you yourself use and develop it a little more fully, the example of Aristotle. I don't want to explore the scholarly issues of the precise relationship between the *Organon* and Aristotle's physical treatises. But you yourself rely to a great extent on the *Organon* and particularly on the *Posterior Analytics*. We have here a model of *epistēmē* ultimately based on a form of *noesis*. Aristotle primarily analyzes scientific demonstration. His remarks about "intuiting essences" are very brief and inadequate; they give rise to all kinds of difficulties. We get just one chapter on this at the end of the *Posterior Analytics*.

In your own terms of goals, methods, and values, I think one could say something like this. What you *really* were showing, taking Aristotle as an example (and one could take others), is that what he takes to be the *goal* of science is to achieve a genuine *epistēmē,* that what he takes to be the *method* of science is induction (in his sense) leading to demonstration, and that his *values* are given by his conception of the world and especially by his conception of necessity.

All of these have radically changed. Let's make the thought-experiment of having someone from the twentieth-century interview Aristotle and say to him: Look, someday we're going to be doing science, what you call science, but it's not going to have the same goals, it's not going to have the same methods, it's not going to have the same values. And then he goes on to describe to Aristotle what science is *now* like in the twentieth-century. Aristotle might well reply: what you call "science" has as much to do with science *(epistēmē)* as football (or an Athenian equivalent) does.

This is, in fact, what the argument of your paper showed. *Retrospectively* we can always tell a story about how science developed, perhaps a good story, one which would justify our giving up of certain goals. It would explain why we modified them or why they changed. What you have done, it seems to me, is to show that anytime we try to point to a constant theme, whether it be predictive accuracy or control or the like, if we want to be historically specific about what these *mean,* we have to admit that they are radically different in different contexts. The challenge to make out the "centralist" position is to show that there *are* enduring substantive goals, methods, values, which are not just verbal ornament. But what you actually *did* in your historical examples was to constantly undermine this.

To go a step further, we could use the same kind of critique that you applied to Kuhn where he says that, roughly speaking, the values of prediction, accuracy, fruitfulness, remain the same. But *we* know, and Kuhn knows, that the meanings and/or beliefs about those concepts radically change in different historical periods. I don't see that you paid cash for the claim you called the "centralist" position. You have to show that there really *is* in some substantive sense a continuity, not necessarily one that makes science something like a natural kind, but one that would allow us to identify enduring aims and methods in science.

McMullin: Not surprisingly, I would say that I have done just that, though not perhaps in the detail needed for persuasion. Enduring does not mean unchanging. Let me return to the example of Aristotle. The disparity between the *Posterior Analytics* and Aristotle's own empirical investigations, particularly in biology, is quite striking and has long been a puzzle for commentators. It is not difficult to recognize in Aristotle's own work, then, a definite tension between the general goals of science formulated in the *Posterior Analytics* and the actual achievement of his biological writings. Many different ways of resolving this tension have been proposed. One that has come in for much discussion lately would suggest that he would have regarded his work in biology as dialectical rather than demonstrative, a stage on the way to *epistēmē* rather than *epistēmē* itself. This kind of tension between stated goals laid down in a formal account of scientific rationality, on the one hand, and the rationality implicit in the actual attempt to achieve those goals is one of the factors that in my view has brought about the development of rationality from one state to the next. This development has itself been a *rational* one; one can specify the reasons why Aristotelian rationality fails. Later writers will point out not only that it fails, but why it fails, even in its own terms; it does *not* lead to the hoped-for *epistēmē*.

When one comes to the seventeenth-century, one still finds attempts, by Descartes and by Newton, for example, to justify the older goal of a definitive, secure knowledge, though in a very different context. But implicit in their work, and explicit in that of others, is a growing realization that the older goal must be modified. Not abandoned, just modified. One will still seek the most secure knowledge available, but one will recognize that it is not "eternal and necessary" as Aristotle had supposed it should be. There is a continuity here that can be traced. There had been a constant effort to secure a knowledge of nature that would satisfy the criteria of *epistēmē*, but already in the medieval period many had noted the difficulties that lay in the way. The actual advances in fields like optics in the seventeenth-century forced people to back off from the claims for *epistēmē* and to settle for something less ambitious.

You mentioned my comments on Kuhn. Kuhn talks about the values involved in theory-assessment, in the light of his work on seventeenth-century science. He does not try to locate these values earlier, but they do emerge, both explicitly and implicitly, in the science of that century. The issue that arises, then, is: what kind of permanence do these values have? There are terms like 'simplicity' or 'accuracy' whose applications in different contexts, and particularly in different sciences, will vary. Nonetheless, they indicate a constraint; that is, if someone either now or in the future were to say: I don't think predictive accuracy in any form is important for scientific theory, the burden of justifying the abandoning of such a criterion would seem heavy indeed. I take it that this is Kuhn's position also. Once it has been discovered in a contingent and empirical way that the observable properties of things may be explained in terms of an underlying structure, we can then say with confidence that there are certain virtues we should expect a theoretical account of that structure to possess. Reliance on these virtues as criteria of "good" theory can in these circumstances be warranted both on historical and on logical grounds.

Bernstein: I still have a question for Alasdair.* It bears on virtue, on the conception of practical reasoning in various historical circumstances. Many people know Alasdair only in terms of *After Virtue,* as though this were his only work. But the MacIntyre I want to speak about goes all the way back to the person who wrote *Marxism and Christianity.* The problem that I find troubling about what you say has nothing to do with relativism or perspectivism; it has to do with *history.* I need to get clear about what *is* your sense of history. I know what it *isn't;* I know what we have to reject in Hegel. But it is ironic that in the course of your work over the last thirty years, you more than anyone have brilliantly shown us that one cannot go back home again. That is the lesson I would have thought we have learned from Hegel. Whatever the glories or the magnificence of the Greek *polis,* whatever the resemblances of later communities to the *polis,* we can't return to the time of the *polis.*

Yet in *After Virtue* to some extent and in your paper here more explicitly, you seem to me to be developing not a Vichian conception of history, but one in which those earlier historical social practices are still an open opportunity. What happened to your earlier understanding of a modernity which in a way destroys, mutilates, or distorts these practices? If you were right, it would seem that while there might be patterns of social practice or principles of practical rationality today that resemble principles of the past, it would be an illusion to think that, in any deep sense, a return to these is possible.

---

*Alasdair MacIntyre's paper, "Practical Rationalities as Forms of Social Structure," which was presented at the original conference, is not included here, by the author's request, since it formed part of a book then in preparation and soon to appear as *Whose Justice? Which Rationality?* (Notre Dame, Ind.: University of Notre Dame Press, 1988).

Rorty: Alasdair's response of "go away!" in answer to a questioner who asked what one should say to a relativist is indeed an appropriate last word. Relativism thought of as a large philosophical view, one more God's-eye view of history or culture, is so silly that all one can do is to say: go away! Relativism couldn't be a coherent philosophical position. It could, as Alasdair said, be expressive of a cultural state of mind. But it would be expressive of a culture in which philosophy, that is, the construction of large philosophical views, was not a prevalent cultural form, or at least, it would be an appropriate expression of a culture in which *that* art form was no longer in vogue. The question that Gary suggested: what's so good about objectivity and truth? goes along with the question: what's so bad about relativism? If you think, as I do, and as Davidson suggests, that it is time to stop analyzing the nature of truth, as well as time to stop talking about objectivity (as Ron Dworkin advised in a recent controversy with Stanley Fish), then it would be time to stop talking about relativism, because objectivity and relativism are two sides of the same coin. What is needed, rather, is an alternative set of terms which might prevent us from rehashing the same issues forever.

Let me say something about my notion of how our culture might, in the end, develop such a set of terms, a set which might enable us to put behind us large philosophical views about, e.g., objectivity, truth, and how to avoid the specter of relativism. I agree with everything Alasdair said about the sequence from the Greek *polis* to the eighteenth-century and onwards to us, except, of course, that I think it is an upward path, a story of success, where he has doubts about that. I particularly liked the sequence from Fielding to George Eliot to Joyce, which I take it even Alasdair thinks is an upbeat story. But what I want to suggest is that one way in which you might characterize the "onward and upward" character of the West is to say that we started out with religion which

was then *"aufgehoben"* into the worship of science, and are
in the process of an *"aufgehoben"* of that too into the worship of art. I think this is a desirable end of the story, or at
least a desirable next stage of the story. To put it in Alasdair's terminology, what we want is a culture worthy of
rootless cosmopolitans; people like Joyce, Nabokov, Emer-
son, Carlyle (and I am tempted to say, Ernan and Alasdair
and myself, since I am not convinced that our roots in
Donegal are quite as firm as Alasdair suggests).

From the point of view of a culture which centered
around art rather than around either religion or science,
rationality would be viewed not as a set of criteria or
structures, but in the way that Alasdair was describing it
in his paper as a set of social practices; or, as I should
prefer to describe it, as the overlapping portions of individual webs of belief and desire. It seems to me that to say
that the rationality of one period is different from that of
another is just a way of saying that the commonly held
noncontroversial beliefs and desires of one period are different from those at another. To look for rationality is to
look for what a community is taking for granted.

I take it that the aim of modern individualistic liberal
culture, the third of Alasdair's stages, the one I like and
that he's not so sure about, is to create a culture in which
this overlap is as thin as possible, one in which the diver-
gences among individuals are as great as possible, as great
as is compatible with social peace. The idea is to arrange
culture so that the only overlap is just enough shared beliefs and desires to ensure tolerance of individuals by each
other. This seems to me the direction in which an upbeat
Deweyan would view the course of recent history as going.
In such a culture, one wouldn't be particularly interested
in the shaping of rationality, because that would simply be
turned over to the historians. They would tell the story
about how *this* overlap rather than *that* overlap occurred
among the various webs that have rewoven themselves in
the course of the past. Conferences in those days would

center around metaphor rather than around rationality.
(This is, indeed, as Wayne Booth has noted, a growing
trend. Booth remarked that by the middle of the next
century there will be more specialists on metaphor than
there are people.) The reason metaphor will be important
is that the ability to speak in a new way will be seen as the
mechanism which makes it possible for people to be, as it
were, their own artist, to shape themselves in a Nietzschean manner. In such an ideal culture centered around
art, the notion of rationality would presumably be replaced by some such notion as metaphor. This is meant as
a comment on those elements in Bernstein's, McCarthy's
and MacIntyre's papers that I disagreed with, namely
those where they exhibit a certain sympathy for Aristotle.
I had no particular disagreement with Gary and Mary in
their papers.

McMullin: In my comment, I want to take off from what
Dick Rorty had to say yesterday about *all* beliefs and desires standing in the same relationship to the world, about
the impossibility of separating *doxa* and *epistēmē*. He attacked the notion that the world is what makes sentences
true, citing Davidson's view that 'true' does not name a
relationship between language and world. Truth can't *explain* the success of science, he said, since this would
make truth a sort of mechanism; one would have to specify *two* sets of wheels. "How things are," again quoting
Davidson, can't explain anything. The relationship between language and world is, as Wittgenstein and Davidson both maintain, causal not representational. But
gravity did not *cause* Newton to acquire the concept of
gravitation. Why, Rorty went on, should we think that
prediction and control are tests of closeness of fit between
concepts and the real? What is so special about prediction
and control? These simply represent human purposes and
are thus themselves to be explained in sociological or psychological terms.

I tried to deal with some of these issues in a historical way in my paper. But I would like to recall and underline one of the points I made. In the seventeenth-century, when the kinds of structural theories with which we have since become familiar arose in natural science, one feature that was not immediately evident at that time was the collapse of any possibility of a primary/secondary distinction. It is curious that just at the point when such a distinction was being undermined, it was being propounded with such force. Nonetheless, it is clear in retrospect that the new theories were introducing terms like 'gravitation' that were precisely *not* related causally to experience. The interpretation of the phenomena of experience by concepts like *gravitation* or *mass* had to be warranted not in terms of abstracting from experience but in terms of the success of the theory employing them. That means that when terms like these are introduced, they are introduced not on *causal* grounds—it is not as though something called gravity causally affected Newton in such a way as to get him to think "gravitation"—but rather on the basis of their furnishing a *theoretical* explanation of the phenomena. The criteria for such an explanation are not just prediction and control; much more important are a certain sort of coherence and fertility of a kind we would *expect* were the theories to be faithful to the structure of the real.

That leads me to say, then, that when Davidson and Rorty say that "how things are" doesn't explain anything, if by 'explain' they mean *causally* explain the concepts of the scientists, they are of course right. To say that it *did* explain in this way would be to eliminate the creativity, the ampliative character, of science entirely. But there is another sense in which "how things are" explains, namely, that it *warrants* these concepts and the theories containing them in the ways in which we have learned to expect good theories to warrant. Retroductive (or abductive) method has been developed in science as an extension of the retroduction we constantly use in everyday life. Its criteria have been sharpened and interrelated in the ways I suggested last night. The warrant, then, for a concept like gravitation or molecule is the success of the corresponding set of theories. And that in turn depends precisely on how things are. Maybe 'precise' is a dangerous term here. I have used the term 'metaphor' in this context myself. But that does not affect the basic point I am making.

Quinn: The last two panelists have introduced the theme of metaphor, which leads naturally to the next panelist, Mary Hesse.

Hesse: What worries me about both Dick and Ernan's comments is that even though they are at opposite poles on the important issues, there is an "asociality" in both their approaches. There would surely be nothing regrettable about devoting conferences to metaphor, but I don't want to take time now on that topic. On the issue of scientific rationality, there is an illicit extrapolation from a notion of common sense, of something shared in common, giving us a common world, to the claim that science has a privileged objectivity just because it is an extension of that common sense. If, when we sit in chairs we expect to be supported, then equally we should accept what science tells us about the world. However, for two different reasons, I don't think that one can make this move. It reminds me of the kind of thing Peter Winch tried to do when he constructed his highly language-game relative account of alternative cultures, bringing together the interests of different cultures in points of crisis in human life, *rites de passage* like birth, marriage, death, which in every society must somehow be recognized. These form common points of concern and interest and commitment and need valid interpretation. They constitute, he suggested, a common practice on which one could build bridges. But this is not anything *like* enough on which to build bridges between radically different cultures. There can be a vast variety both of sociological features and of cultural and cosmological belief systems, all of them consistent with these basically biological features.

Another type of discussion, a debate of the Davidson-Dummett kind, about whether we can talk about the truth of most of our propositions, seems to me to rest upon a similar confusion, in that Davidson is assuming that most of our propositions are going to be the sort of things that our common sense requires us to talk about. That this is a rabbit, that we can shoot the rabbit if we go about it the right way and that eating the rabbit responds to human needs, is taken to be typical of the kinds of ways in which we understand realism and rationality. It seems to me that this again is much too narrow a basis to provide the conceptual frameworks with something in common, if I may dare to put it that way when talking about Davidson. It is far too narrow a basis in itself.

Moreover, the notion that science somehow extrapolates in an absolutely common way and therefore gives a common objective world of common natures seems to me to be not true for all the reasons I have given. But even if one takes it that in our culture it *is* true because we have, as it were, constructed an objectivized world, not just a complex version of what science goes on to tell us on the narrow basis of common sense, this is quite clearly not the only way one could go about it. It is a peculiarly historical phenomenon of roughly, I suppose, the last three hundred years. Aristotle's understanding of science is not the same as this one. This leaves me with two particular historical problems; it seems to me that there could be a great deal of illumination if historians of science would address themselves specifically to them. How did this notion of an objectivized scientific world get started? And how did it gain such power over the human imagination? When and how did it happen? Maybe it happened somewhere in the late sixteenth and the seventeenth-centuries, first of all in relation to the sudden decision (I take 'decision' here in a social sense) that precision of measurement is to be treated as important, that Kepler's eight minutes of arc suddenly becomes a crucial sort of criterion for what a cosmology should offer. The astronomers of previous centuries, including the Greek ones, would not have cared about eight minutes of arc. An interesting historical question, which has, of course, been partially studied, is why this should have happened when it did. Kepler relied on the new instruments, of course, but were they cause or effect?

The other type of historical question is one that arose in my mind after Dick Rorty's talk: How did this, what I call "ideology" of the truth and objectivity of science get started? It seems to me that these notions become different somewhere around the first half of the seventeenth-century. They come to be dominated by the kinds of things the astronomers and the corpuscular mechanists were then trying to do. There is no doubt that *they* were realists, that they saw this as a tremendously powerful social instrument, as well as a natural instrument. But I would like to know why this happened. Was it a sudden illumination, as some suggest? Even a slight degree of historical sophistication would suggest that this is not the right interpretation. We need much more detailed analysis. If we get this, of course, it doesn't show anything philosophical, but it might indicate that philosophers have been worried about what are, in fact, rather temporary historical features of a particular culture.

McCarthy: I want to make just a few brief remarks on what others have said. First, if we had spent more time discussing social-scientific rationality as part of the picture of scientific rationality generally, we would have had a very different discussion. Our discussion here has been based on a shared view of scientific rationality drawn from paradigm cases in the natural sciences. If we had talked about social sciences, the paradigms of rational practice would be widely divergent even within the same discipline, and there would be no agreement about how to go about inquiry. One would find a whole range of rational practices, some looking like textual interpretation of historical narratives, others trying to look as much as possible like natural-scientific rationality. It is clear that there's no good reason for drawing the line there; if we're interested in scientific rationality as part of the more general question of the nature of rationality, especially in the context of the sort of social and cultural problems we have been discussing, it would be important to take into account as broad a spectrum of the types of rationality as we can.

Truth and objectivity are not just private possessions of the sciences. Questions concerning them did not begin in the seventeenth-century. One could find a coherent and continuous story of changing argument and counter-argument from Aristotle to the present. Truth and objectivity are claimed by epistemologists and metaphysicians and dialecticians and everyone else who perceive themselves as participating in the philosophical discourse, at least as that has been traditionally conceived. That is why I think that Dick Rorty's idea that it might be time to worry about something other than truth and objectivity and relativism, and to start talking about something else, raises such fundamental questions.

It has become fairly obvious to me, as it also has to Dick Rorty, that you can't make sense of philosophical practice without invoking those notions. One can very easily show that the people who challenge them, including Foucault, for example, and indeed Dick himself, are caught in a difficulty since insofar as one participates in philosophical discourse, one must implicitly appeal to *some* notions of truth and objectivity. Dick has recently, and quite consistently, responded by saying: Well, that is more or less the case; it is hard to participate in philosophical discourse without subscribing to a whole set of rules and therefore implicitly undermining one's own case. So the thing to do is to stop talking in this way and go on to do something else, and show that that something else is more interesting to do.

That raises two thoughts in my mind. One is that, under present circumstances at least, one could look at it in terms of market conditions. What will be the most interesting approach? How are people to be attracted to it? Who will have most access to the mass media or to government funds or to large conference programs? I don't know how exactly it will work out there. But if one generalizes at a social level, the way that Dick did in his remarks, talking about the *Aufhebung* of science in a broad sense into art, then I really have my doubts. That kind of remark is, I think, based on a rather unwarranted optimism. I have a feeling that if there is an *Aufhebung* of truth and objectivity, we are not going to get a culture of independent artists among whom everyone has power; we are going to get just exactly the opposite. Truth and objectivity have had, besides the negative aspects that we have been focussing on, the positive effect of challenging precisely the kind of thing that Dick most dislikes in our culture . . . ideology and the like. I am not optimistic, if we stop talking about such things as truth and start talking about art instead, that things will turn out better.

Indeed, if one looks at traditional attempts to aestheticize social and political discourse, they have turned out to be on the whole disastrous. I don't know why we should be more optimistic about the future. I think Dick is taking for granted much too much: the social and political framework of justice within which the "rootless creative cosmopolitans" could have their due. Of course, the really difficult question is: how do you create and sustain a social and political framework of justice, equality, freedom, where people can lead their lives in the desired way? I don't think that by getting rid of the discourse of truth and objectivity, you will be able to do that.

Quinn: Before Dick Rorty responds, there are several challenges facing Alasdair MacIntyre. Among other things, he has been accused of taking the relativists too seriously, and on the other hand not scaring them enough.

MacIntyre: I am not sure what I am going to say in response to *that* challenge. I have basically four remarks to make. First is a footnote to a discussion between the other participants in which I have not so far taken part. I don't believe that it is ever useful to talk about realism or alternatives to realism. When one speaks about anything in the terms that lead one to invoke realism, one should always talk about realism only with respect to a particular class of entities. It seems to me that there were moments within the history of science when it was important that realism in regard to a class of entities was part of the scientific theory itself, part of the science, and not in the least a philosopher's position about science *ab extra.* The controversies about realism should be understood as part of the controversies that belong *within* science and not at all as controversies as to how science is to be construed by philosophers. Once one sees this, a great many of the problems disappear. I don't think that there was a single *unconditional* realist in the world before Hilary Putnam, who proposed this thesis only in order to attack it afterwards.

Secondly, we have been considering the outcome of a number of trends in the last twenty-five years, to which Kuhn, Foucault, and many other people have contributed, including indeed some of the people sitting at this table. What we have been considering for the most part is whether or not these trends have fatally undermined the applications of certain notions of truth and rationality and objectivity to, and within, the sciences. Here I simply want to put in a plea again for a distinction made by Mary Hesse (though I might sometimes disagree with her

applications of it). I think that the notions of truth and
rationality and objectivity were deployed not only as limited notions *within* the sciences but as part of an ideology
of the sciences which has largely dominated our culture
since the Enlightenment.

The effect of the critiques made by Kuhn, Foucault,
Derrida, and the others has been to dissolve the ideology.
I do not think that it should lead us to withdraw the
application of the more precise and limited notions
within the sciences. That is why I agree with Ernan, in
large part. But I do think that the collapse of these notions has meant the collapse of the arena of public discussion in important ways. I think the differences between us
here are really fundamental and thoroughgoing, and that
if we push them much further than we have done at this
conference, we would start finding it very difficult to talk
to each other. We would have to listen carefully to Tom
McCarthy's prescriptions for the sort of dialogue between
cultures that we would then be engaged in. The modern
university fosters the illusion that anyone can talk to anyone else about anything. It's an illusion that ought now be
given a decent burial.

Let me now turn to Dick Bernstein's challenge. Because
I shall be very brief, I shall be inadequate anyway, but I
suspect that if I were to speak at great length, I would be
seen as even more inadequate. Dick says, in effect: you at
one stage knew enough history to know that modernity is
here irremediably, and there is nothing to be done about
it, and therefore anything that one says has to, as it were,
take as its starting point the positions of modernity as it
now is. Well, I think he has misread my past. The last
questioner in the discussion after my paper said to me
afterwards that he was really requesting my autobiography
when he asked why I dismissed relativism. And I said:
well, I guessed that, but since the answer to your question
"Why do you respond as you do to relativism?" would,
have been, in part, "Because of my aunts," this would

have been rather inappropriate in a general philosophical discussion, though it would have been a true answer.

In fact, my intellectual history is the story of the struggle between my aunts and Karl Marx. My aunts and Karl Marx began with a large measure of agreement. All three of them deplored liberal individualism, root and branch. And it has never occurred to me in consequence that there was *anything* to be said for liberal individualism. I may have changed my mind on other things but never on that!

What led me to take modernity seriously was that for a long time I believed that Marx had provided essentially the right framework for understanding it. And after I stopped believing that one could put it in precise Marxist terms, I went on much longer thinking of it in what were essentially Hegelian terms. It was when I saw that none of that would do, that things were much worse than could be allowed within either a Marxist or a Hegelian framework that I realized that in many ways they were also much better.

What I mean by that is this. The collapse of modernity, which we have been seeing in the collapse of those notions of truth, rationality, and objectivity, certainly isn't going to usher in a period when we congregate around the arts and talk about metaphor. What it *is* going to do is to make the key problems of modern society appear as insoluble as in fact they have been for a long time, but I hope much more clearly. And I think that if one asks what it is that leads one, what standard was always implicit in understanding the problems, what standard it is in the light of which we are able to understand the impasse and the incoherences of modernity, it turns out to be the case that all the time we have been presupposing Aristotle. It is coming out very interestingly now, for example, in work being done on Marx by a number of people, like Carol Gould and Harry Lubasz, that Aristotelian standards have been in the background of our culture to a much deeper and a more

lasting extent than we had supposed. What I think happens now is a confrontation between the resources provided by that element from our past and what is actually
appearing yet. So Dick and I really do disagree in a way
that could never be resolved in a discussion such as this
about the character of the modern age. He looks forward
to the arts as providing the rhetoric, the encounters, the
creative activity of the immediate future; I look forward to
the day when the last proponent of the "strong program"
in the sociology of science is strangled in the entrails of
the last expert in the theory of metaphor!

Rorty: I agree that you can cite all sorts of horrible examples of the disastrous effects of the aestheticization
of social discourse. But you can cite as many or more examples of the disastrous effects of "scientized" or "theo-
logized" social discourse. It seems to me that the track
record of ideological pushes to change the character of
our social discourse is uniformly bad. I would like to
think that by slow, insensible Deweyan transitions, given
the miraculous preservation of the institutions of the free
world, over the centuries the rhetoric of public life
might shift from a rhetoric of science to a rhetoric of
poetry. But I agree that to try to make an explicit ideology
of it would make things even worse than they are now.

  Let me say something about Alasdair's distinction between the use of 'true' and 'objective' within the ideology
of science and a more precise and limited use within science itself. I agree with him and with Ernan in regard to
the way the terms 'true', 'objective', and 'really there' are
used within science itself; nobody would have any com-
plaints about this. But Alasdair regards that use as more
precise and limited. I regard it as trivial because noncontroversial. That is, I am not clear what a scientist who
says: "I regard them as real and I'm going to look for their
internal structure" is supposed to do about a colleague
across the way who thinks of them as heuristic fictions. I

am not sure what, after they have decided what research program they are going to pursue, their particular philosophical glosses do for them. I'm pretty sure, as I think Alasdair is too, that trying to connect this use of those terms with the use made by Hilary Putnam, for example, is not going to do the scientist or anyone else any good.

Quinn: One thing that Gary Gutting remarked on was a sort of historicist turn in science studies in the last couple of decades. We are privileged to have with us Peter Hempel who has been active throughout the entire period as well as having shaped its prehistory. It would be appropriate for him, if he wishes, to comment on the central ideas before us.

Hempel: Let me begin with a personal remark. I have found these discussions extremely stimulating. I was taken aback by some of the ideas, especially by the idea of an impossible future where people sit around exchanging metaphors. This seems to be an extreme case of "culture fiction." To reach such a stage and in order to know whether it would make people happy, I would say, with Tom McCarthy, that we would need procedures which are rational roughly in the way in which scientific inquiry has been for a very long time, in the way that Tom Kuhn, for example, has described it. Whatever we might want to say about whether people should *want* such a future, it would have to be brought about in ways that are scientifically rational, in this broad sense.

I too have in the past been puzzled by what exactly could be meant by statements about "reality," about the "reality" of objects. In science we can ask about the objects that are postulated by a particular theory of elementary particles, for example. There I would say that since the theory says that matter consists of such particles, whatever grounds we have for believing the theory, we have the same grounds for believing that those things exist. If one wants to say that they are "real," one can certainly do so, but this doesn't add anything. There is no transcendental sense of "really real." And of course since our grounds for making such an assumption may change, one may expect that our views as to what there is, that is as to what is real, may change.

About truth, I think there is something analogous. If we have any grounds for proposing certain theories, we have exactly the same grounds for asserting that those theories are *true*. To say, at least in the traditional semantical sense, that a statement is true is simply to say that what it asserts is the case. And so there is no difference except in semantical level between the two assertions. Whatever support one of them has will count for the other too. This support can always be called into question, of course; in fact we expect that it *will* be called into question. There will never be a final point in this evolution; I don't think there is a limit, or at least I don't know how to make this notion of a limit clear.

I have gone through an evolution on these matters; it was fairly painful for me but this belongs to autobiography. Reichenbach would have said: Don't reminisce or reflect on how you felt about this! I found it illuminating eventually, after I had been exposed to the shock for quite some time, to be made aware of the cultural and historical aspects of the development of scientific thinking. This has been very liberating. I don't know my stand on every particular detail that might arise but I certainly think that one cannot go so far as to envisage, except perhaps in a logical sense as a possible world, a situation where there would be no interest in scientific thinking at all. There are biological reasons for expecting that there will *always* be an interest, in order to improve our accommodation with the world. Therefore I am more optimistic about this than some of the other speakers seem to be or than their comments on ideology seem to indicate, at least. But I have learned a great deal here, and have only wanted at this point to indicate the difficulties I have felt about some of the more extreme implications of these new views.

Questioner: I would like to direct a question to Mary Hesse about her remarks on the "strong program." She suggested that in the development of this program there was a fundamental distinction between *accepted* beliefs and the traditional analysis of knowledge in terms of *justified true* belief. It seems to me that if this is an assumption of the strong program it amounts to accepting a fact-value distinction. It is as if one could identify our currently accepted beliefs without reference to justification or truth—that is, to an ideal of what knowledge should be. It seems to me that one of the ways in which the strong program fails is that such an identification is from the beginning impossible.

Hesse: Yes, of course, the defenders of the strong program *would* insist on this distinction, since they would say that to claim that we know which are the justified true beliefs would involve us in a circularity. But the "program" of the strong program would be to discover in social terms which beliefs are *taken* to be justified and true. There are various ways in which one could go about this. And objections to this approach would have to be answered one by one. One objection, for example, is that in presenting their thesis and the evidence for it, the proponents of the strong program are themselves making claims about the historical data and drawing inferences about social practice and so on. But this is the standard self-reflexive objection, which they can respond to by saying: we are living in a culture where we can delineate as observers, just as we could in an alien culture, what is the general, accepted consensus as to what counts as knowledge and what the modes of inference are that are generally regarded as rational in our culture. And we are going to use *those* modes ourselves, with no further question about justification.

Justification in any deeper sense does not arise at either level, according to them.

The analysis in terms of justified true belief is part of the story told in our culture about what is going on, but one has to see this as one exercise, among many, of anthropologists and sociologists and historians of ideas as they approach alien cultures and notice that something called "cognitive belief-systems" occur in them. They have, of course, to decide what is to count as such a system, but supposing that this problem of hermeneutics and sociology can be solved, then they can use this same methodology to apply to our own society to find out what can be delineated there. And the practice of constructing an ideal type of what "knowledge" is or what the alien culture's "knowledge" is, is an important feature of what they find.

McMullin: To an extent that I had not anticipated when this conference was first planned, our deliberations here have brought into focus some very deep divisions regarding the present and likely future status of the notion of rationality, whether scientific or not. Alasdair's remark that if we were to press the differences between us here much further than we have, a great silence would descend, is surely an apt one. It might become difficult, he suggested, to find ways of dealing with the differences other than to say "Go away!" I am not going to close our discussion, needless to say, by exclaiming "Go away!" Instead, I will suggest that the conference has shown how people who deeply disagree on fundamental philosophical issues can still remain colleagues and treasured friends over many years.

NAME INDEX

Adorno, Theodor, 191, 194–95, 201–
  6, 209, 214, 216–19
Arbib, Michael, 121
Aristotle, 2, 26–27, 30–31, 53–56,
  189, 212, 225, 226–28, 235,
  237, 241
Aron, Raymond, 153
Austin, John L., 176
Averroes, 31

Bachelard, Gaston, 158–69, 178, 184,
  187
Bacon, Francis, 2, 32, 35, 50
Baker, Keith M., 217
Barnes, Barry, 12, 45, 77–78, 83,
  90, 93–94, 120
Benjamin, Walter, 202, 214, 216,
  221
Berkeley, George, 68, 74
Berlin, Isaiah, 214, 220
Bernstein, Richard, vii, 232, 234,
  240
Binswanger, Ludwig, 158, 187
Black, Max, 165
Blondlot, M. R., 114
Bloor, David, 13, 45, 77–78, 83, 90,
  93–94, 100–1, 120
Blumenberg, Hans, 74
Bohr, Niels, 316, 333
Bonald, de, Louis G. A., 217
BonJour, Laurence, 151
Booth, Wayne, 232
Boyd, Richard, 59–60, 73

Brooke, John, 122
Burger, Ronna, 220

Canguilhem, George, 158–59, 164,
  166–67, 178, 184, 187–88
Carlyle, Thomas, 231
Carnap, Rudolf, 49
Cavailles, Ferdinand, 158
Chomsky, Noam, 73
Chubin, Daryl, 120
Collins, Harry, 105, 111–21
Comte, Auguste, 108, 157, 169
Condorcet, Antoine Marquis de,
  191–98, 206, 217
Copernicus, 31, 44
Costa de Beauregard, Oliver, 47
Cushing, James, 47

Dallmayr, Fred, 76
Darwin, Charles, 76
Davidson, Donald, 55–56, 68–69,
  72, 74, 91, 95, 149, 230–35
de Beauvoir, Simone, 153, 186
de Saussure, Fernand, 174
Democritus, 19
Derrida, Jacques, 213–16, 220–21,
  240
Descartes, Rene, 2, 32–36, 149,
  159, 227
Dewey, John, 50, 69–70, 74, 211,
  213, 220
Dilthey, Wilhelm, 100

247